SMALL-SCALE FOOD PROCESSING

Small-Scale Food Processing

A guide to appropriate equipment

PETER FELLOWS and ANN HAMPTON

INTERMEDIATE TECHNOLOGY PUBLICATIONS
in association with CTA • 1992

Published by Intermediate Technology Publications
103–105 Southampton Row, London WC1B 4HH, UK.

© Intermediate Technology Publications 1992

ISBN 1 85339 108 5

A CIP catalogue record for this book is available from the British Library

Typeset by Inforum, Rowlands Castle, Hampshire
Printed by SRP Exeter

Contents

Preface

The level of interest in small-scale food processing in developing countries has increased dramatically in recent years. It is due in part to the promotion of income-generating activities to help increase incomes and employment in rural areas, and also in part to the success of agricultural development programmes in some countries which have produced food surpluses then requiring preservation and processing.

As a result there has been a corresponding upsurge in enquiries about food processing, and especially about the availability of low cost, small scale equipment and where it can be found.

In response to this need, ITDG began in 1989 to produce this guide. It is based on the highly successful *Tools for Agriculture* guide which was first published in 1965, updated in 1985 and now re-published as a companion volume to *Small-Scale Food Processing*.

One difference between agricultural processes and food processing is the much longer time intervals between different stages of production in agriculture. For example, there may be several months between the stages of ploughing, weeding and harvesting. In food processing the intervals between stages (or unit operations) in a process is much shorter, often a few minutes or hours. It is therefore necessary to select equipment for a particular stage that has a similar throughput to other pieces of equipment used in stages that precede and follow it. Delays in the process may otherwise be caused by one piece of equipment that is too small, and food may consequently spoil.

In food processing it is therefore necessary to look at the *whole* process when deciding on the equipment required, and for this reason we have included in the first part of this book chapters that describe the stages and equipment needed to process selected foods in each commodity group.

The second part catalogues the different sizes and types of equipment.

The selection of equipment to be included in the catalogue section was not easy and we make no claim to be comprehensive. The overriding principle was that the equipment should be suitable for low or medium-income producers in developing countries. So you will not find the complex, automatic, continuous equipment used by high-technology food processors in industrialized countries.

However, the scale of production required by small producers and the availability of services (electricity, clean water, gas, servicing and maintenance facilities etc.) will obviously vary in different countries and between regions of the same country. Similarly the amount of money available to invest will also vary considerably. We have therefore included, wherever possible, a range of equipment from simple hand-operated tools which are often suitable for local manufacture, through to larger factory-produced machines that require the above services, but are still small-scale by industrial standards.

One of our main aims has been to include equipment manufactured in developing countries to promote direct South-to-South trade relations. It should be noted that the response to our request for information was overwhelming from manufacturers in India, but elsewhere it was modest, particularly from African countries. We would be grateful to hear from manufacturers who are not included in the guide in order to improve our database for future editions.

We hope that this first edition of *Small-Scale Food Processing* will form a useful addition to the relatively limited information available on the subject and serve its purpose as a tool to assist the improvement in peoples' livelihoods throughout the developing world.

PETER FELLOWS

Introduction

Every parent in the world knows how to process foods – they do it every day when preparing meals to feed their families. But 'food processing' as a scientific and technological activity covers a broader area than food preparation and cooking. It involves the application of scientific principles to slow down or stop the natural processes of food decay caused by micro-organisms, enzymes in the food, or environmental factors such as heat and sunlight, and so preserve the food. Food processing also uses science together with the creative imagination of the processor to change the eating quality of foods and provide people with interesting variety in their diets (as seen in the hundreds of different types of foods such as pickles, cheeses and biscuits that individual processors make throughout the world).

So the reasons for processing foods may vary. In some countries the main aim is to preserve basic foods such as cereals, root crops or vegetables against periods of shortage; i.e. to increase the food security of populations throughout the year.

Others wish to process foods to create employment and to generate additional income, either as a little extra money to supplement family incomes or to establish and expand a recognized food processing business. Small-scale food producers often start by working from home using domestic equipment; they often have little money to invest in equipment and little access to credit. However they must be able to produce uniform quality foods under hygienic conditions.

Some people will see food processing as their main source of income. They are entrepreneurs who will take out a loan to buy specialized equipment and secure working capital and, if successful, they will develop business and marketing skills to expand and diversify their enterprise.

Food processing as a business

For those who earn an income from food processing there are special problems which make this type of business different to most others: many raw materials are highly perishable and will spoil quickly after harvest or slaughter unless properly processed. They may also be highly seasonal which means that they can only be processed for part of the year, or alternatively they are part-processed for intermediate storage until they are needed.

Added to this, foods are biological materials whose composition varies as a result of the actions of weather, pests and diseases. This can mean unpredictable supplies and costs for raw materials. Some processed foods also have a seasonal demand (such as for festivals and ceremonies), which further complicates the business of food processing.

Even after processing, foods do not keep indefinitely. The 'shelf-life' of processed foods can vary from a few days to several months or years. The distribution and sales methods used by the processor must be suited to the expected shelf-life of the food and carefully organized so that customers receive the food before it spoils.

Packaging is an important means of controlling the shelf-life of foods, but there are universal problems in finding suitable packaging materials in developing countries. This is one of the most important constraints on small-scale food processors. The technically advanced plastic films, cartons and cans usually have to be imported, and even if foreign exchange restrictions or available supplies permit, they can be very expensive. Traditional alternatives such as leaves, clay pots etc. do not perform as well technically and are often perceived by customers as inferior. This puts the business at a marketing disadvantage compared to equivalent imported products.

In all food processing activities there is the overriding concern to avoid food poisoning. Food is the only commodity that people buy every day and take into their bodies. Processors and processing methods must meet strict standards of cleanliness and production control to avoid the risk of harming or even killing their customers by allowing the growth of food poisoning organisms in their products.

In no other type of business do processors operate under these multiple complex technical constraints. However, they do share with other types of small business the difficulties of operating in a frequently hostile economic environment. In the majority of developing countries the bulk of food processing enterprises are on a small scale and are located in the informal sector. They are rarely formed into associations and have little economic power or ability to seek such assistance as may be available. They will often need intermediaries, such as extension agents, to guide them to appropriate solutions for their own individual problems.

The larger, formal food processing sector may receive government support in the form of subsidies, foreign exchange allowances, price stabilization or guarantees and access to specialist advice. In contrast, the small-scale informal sector has no political influence, despite its combined voting power, and is therefore subject to the vagaries of the national (and often international) economic climate.

Support to small-scale processors

Despite these combined problems, many governments and development agencies are promoting food processing as a means of alleviating poverty in rural and peri-urban areas. The reasons are not hard to find: food is familiar to the target groups, the raw materials are readily available (often in surplus), the technology is suitable for small-scale operation and is accessible and affordable, equipment can often be manufactured locally, creating further employment, and the products, if chosen correctly, have a widespread demand. Compared with some other technologies, small-scale food processing is particularly suitable for women and it has few negative environmental effects.

The selection of suitable products for small-scale manufacture, and then the process by which to make them, requires very careful consideration. It is not sufficient to assume, as many 'advisers' do, that simply because there is a surplus of a raw material each year that a viable food processing venture can be created to use up the excess. There must be a demand for the processed food which has been clearly identified before a process is set up. Otherwise the most likely result is to produce a processed commodity that no one wants to buy and substantial financial losses to those involved.

In general the types of products that are suitable for small-scale production are those for which a high value can be added by processing. Typically, cereals, fruits, vegetables and root crops have a low price when in their raw state, but can be processed into a range of baked goods, snack foods, dried foods, juices, pickles, chutneys etc. which have a considerably higher value. The high added-value means that the amount of food that must be processed is relatively small, and hence the size and type of equipment required can be kept at affordable levels.

A second general statement, which is particularly relevant for those with little previous experience of food processing, is that the product selected should have minimal inherent risk of food poisoning. Acidic foods (such as yoghurt, pickles, fruit juices, jams) and most types of dried foods have a low risk of transmitting food poisoning micro-organisms. In contrast low-acid foods such as meat, milk, fish and some vegetable products are much more susceptible to transmitting food-borne illness through poor hygiene of workers or incorrect processing conditions.

Some types of process have a larger inherent risk of causing food poisoning than others. Canning, bottling, chilling or freezing of low-acid foods are each more risky than, for example, acid fermentations, drying or concentrating foods. In addition some processes are much more expensive to set up and operate than others: for example, canning and freezing are more expensive than drying, bottling (e.g. fruit juices) and baking at a small scale.

However an individual producer should not base a decision to produce a food on availability of raw materials, cost of equipment and risk of food poisoning alone. He or she should conduct local market surveys to find out which processed foods are in demand and how much people will pay for them throughout the year. The scale of production is then set to meet a pre-determined proportion of this demand. From this scale of production, together with technical advice on the best way to process the food, it can be decided what size and type of equipment is required. One function of this guide is to show what equipment is currently available and the approximate price of each piece. These costs are needed to calculate the approximate capital required for equipment (and if necessary the loan required) in a feasibility study of any proposed production before getting precise details from equipment manufacturers.

Another function of this guide is to prompt ideas for local adaptations of equipment to meet particular needs. If a processor has a problem in finding suitable equipment it is not necessary to 're-invent the wheel' to find a solution, but similarly it is not always necessary to go to the expense and time involved in importing equipment. It is hoped that this book can give ideas from which a processor and a perceptive workshop owner can produce equipment that will provide an appropriate answer to the problem.

We recognize that knowledge of these appropriate technologies alone will not ensure their adoption. Proven prototypes may be needed for demonstration and those who are convinced of the effectiveness of a technology may need financial support to acquire and promote it. This in turn may require the collaboration of national food research institutes and university food science and technology departments for development and testing of technologies for local needs.

However, knowledge is a pre-requisite to such development and it is hoped that this publication will have a role to play in disseminating information about food processing technologies.

Appropriate food technology implies affordable, locally produced and locally repaired, reliable technology that has a suitable scale and complexity of operation for the people who will operate it. It will help increase incomes and improve (or at least avoid worsening) income distribution.

In practice, however, it is likely that the consequences of introducing a new technology are largely unpredictable. It is true that potential adverse effects of a new technology on poor producers can be predicted to some extent and therefore avoided by careful studies before a project is implemented. But the large number of factors that are in play during a technological change prevent an accurate prediction of the final outcome and of who will benefit. There is therefore a need for sensitivity and understanding of the social and cultural context in which the introduction is planned.

If a demand exists for the products of technologies described in this guide, the changes will come. It is the concern of implementing agencies to try to ensure that the change benefits disadvantaged groups rather than further threatening their livelihoods.

We are aware that it is the staff of implementing agencies who will read this book and not the ultimate intended beneficiaries – the small-scale food processor. There is therefore a responsibility to evaluate carefully the technologies described to ensure their effectiveness for each individual processor who is being assisted.

The criteria that will help in deciding whether to recommend a technology are complex and inter-related but are likely to include the following:

○ technical effectiveness (whether the equipment will do the job required at the indicated scale of production),
○ relative costs for both purchase and maintenance of equipment and any ancillary services required,
○ operating costs and overall financial profitability,
○ health and safety features,
○ conformity with existing administrative or production conditions,
○ social effects such as displacement of a workforce,
○ training and skill levels required for operation, maintenance and repairs,

Introduction

○ environmental impact such as pollution of air or local waterways,
○ flexibility to perform more than one function,
○ compatibility with other parts of a process.

However, it must be stressed that each of these factors is an aid to judgement by staff on the spot and not simply a checklist. Each will have a different weighting in different circumstances and there can be no simple solution to the difficult task of weighing up all factors in a particular situation and making the 'best-fit' from the available technologies.

The technologies and equipment described in this book will undoubtedly affect the economic status of many people, and not always positively. However, in comparison with large-scale, automated technologies used by food processors in industrialized countries, those presented here are relatively benign. The loss in productivity by using predominantly manual procedures is insignificant compare to the under-utilization and high investment costs of larger automated equipment. The gain in employment and sparing use of resources makes these technologies more sustainable and therefore more valuable to the small-scale processor and ultimately to the national economy.

How to use this book

The guide is intended for use by the following people:

○ development workers who wish to purchase of equipment in order to set up or expand a food processing unit;

○ advisers, government officials or development agency staff who wish to know the kind of equipment needed to process particular types of food;

○ those who wish to know the range of small-scale food processing equipment currently available.

The guide can be used in the following ways:

○ to find out what equipment is needed to process a particular food, and the approximate cost;

○ to find the name and address of a manufacturer for each piece of equipment required;

○ to learn more about the potential uses of various equipment.

The guide is arranged in two parts:

Part One **Food processes** (chapters detailing the production stages and equipment needed to process specific types of food).

Part Two **Directory** (details of individual items of equipment and suppliers).

In Part One the food processes are arranged in chapters covering different food groups or types of product (e.g. dairy products, beverages). Each chapter has a description of the preservation principles for the commodities, some observations on suitability for small-scale processing and a description of the process and the equipment required. Reference numbers link each item of equipment to the entries in the directory.

In Part Two, the directory is arranged in 68 sections, in each of which the main features of equipment are described and the range of equipment available is listed. The sections are sub-divided where appropriate into manual and powered equipment. In most sections a photograph or line drawing is used to illustrate an example of the type of equipment described.

It is important to recognize that illustrated examples have been selected as representative of the equipment being described and this does not imply superiority in any way over similar equipment from other manufacturers.

The names and addresses of manufacturers are listed below each group of equipment, and are also indexed by country at the back of the book.

If you wish to know the type of equipment needed to produce a particular food:

○ look up the food on the contents page or index

○ turn to that page in Part One and see the list of equipment in the right-hand column of the relevant table.

Equipment required

Processing stage	Equipment	Section reference
Prepare raw material	Fruit and vegetable cleaners	14.1
	Peeling machinery	51.0
	De-stoners	21.0

If you want to know where to buy the equipment and its approximate cost:

○ make a note of the section numbers from the table

○ look up each section in turn in the Directory to find the names and addresses of suppliers and the price code for the equipment.

51.1 Peeling equipment

Manual peeling machines

FRUIT PEELER

Fruit is positioned on a central pivot. As the handle is turned the fruit is revolved over a stationary knife and the peel is removed.

PRICE CODE: 1

MAQUINARIAS Y AUTOPARTES EL LATINO
Av Grau No 1191
LIMA
PERU

NOTE: In many cases a range of equipment will be shown and it is necessary for you to select the most appropriate type and size for your needs.

The price codes used throughout this guide are as follows:

Price code	Cost ($US)
1	0 – 170
2	171 – 850
3	851 – 1700
4	>1700

If you want to know which manufacturers supply equipment in a particular country:

○ look up the manufacturers' index and find the country (in alphabetical order)

How to use this book

○ look up the manufacturers' name and address (also in alphabetical order).

When you have found the equipment you need, we suggest that you write directly to the manufacturer for an up-to-date specification, current FOB price of delivery to your country, delivery times and cost of spare parts. Remember, it will help the manufacturer to reply quickly if you include the full name of the equipment and model number where available. The more information that you can give about your needs, product throughput, voltage and so on, the better the manufacturer can provide you with the right tools for the job.

We have made every attempt to ensure that the information in this guide is accurate but inevitably changes will have taken place since compilation. We ask you to contact us with any corrected information you are given so that we may update our records for future editions.

Please write to:

The Manager, Small-Scale Food Processing Publication, ITDG, Myson House, Railway Terrace, Rugby CV21 3HT, UK.

It must be stressed that this guide relies on information supplied by manufacturers and that inclusion of an item of equipment is not a recommendation by ITDG, or a guarantee of its suitability or performance.

Whilst every effort is made to ensure the accuracy of data in this guide, the publishers and compilers cannot accept responsibility for any errors which may have occurred.

Please note that equipment availability and specifications are subject to change without notice and these should be confirmed with suppliers when making enquiries or placing orders.

Improvements to this book

Small-Scale Food Processing is only as good as the information provided by manufacturers. You can help us to improve the quality of this guide by doing any of the following:

1. Please fill in the short questionnaire at the back of this book and send it to us. This is most important as it will tell us what you want from the guide.

2. If you use the book to contact a manufacturer, please tell them that you found the address in *Small-Scale Food Processing*. They will then be more likely to supply us with information next time round.

3. Please send us details of any manufacturers of appropriate equipment that do not appear in this edition (name, address and, if possible, equipment brochures).

4. Please write with any suggestions, criticisms or comments you have on the book.

5. Please tell us of any errors you find in this book.

6. Please tell us of any sources of information that we should be using to increase our coverage of manufacture.

It is only through your help that this unique information resource can be improved to provide you, the user, with what you require.

Acknowledgements

This guide could not have been compiled without the efforts of many people. Thanks are due to the many company managers and directors around the world, who took trouble to reply to our letters and without whose information, photographs and diagrams there would be nothing to publish; also to the companies that responded but whose equipment was outside the scope of the guide. I would particularly like to thank the following people who have made a major contribution to the successful completion of this book: Ruth Fairman for organizing the equipment classification, the mailings to the companies and compiling the computer database; Ann Hampton for organizing the later stages of production, including editing and organizing the technical information, preparation of the introductory texts and commissioning artwork.

Thanks are also due to Jeremy Herklotz, Tim Ogborn and Richard Holloway for advice and guidance, to Matthew Whitton for preparing the artwork, to Jean Long of ITDG's photo library and to the FAO for photographs, to our colleagues at the *Groupe de Recherche et d'Echanges Technologiques* (GRET) for sharing their technical information, to Paul Senescall for technical editing, to Ros Patching for editing the manuscript, and to other staff in ITDG for their support and assistance throughout.

Finally I am particularly grateful to those who provided financial support for the preparation and publication of the guide, notably the Technical Centre for Agricultural and Rural Co-operation (CTA), the Overseas Aid Committee of the States of Jersey and the Overseas Development Administration of the British Government.

P.J.F.

PART ONE

FOOD PROCESSES

1. Fruit and vegetable products

Many organizations, from both the government and non-government sectors, are actively promoting the processing of fruit and vegetables. The reasons for this include:

○ Attempts to preserve seasonal gluts which often lie rotting on the roadside.

○ Difficulties in storing large quantities of fresh produce without incurring heavy losses.

○ Small local markets for the large quantities of fresh produce in season.

○ Ineffective distribution and transportation to meet demand in other, often urban, areas.

Due to the above constraints, rural producers are often forced to give produce away or let it rot. To prevent this loss, many may be tempted to convert such gluts into value-added products to be sold in the urban areas. However, before production begins, it is advisable that the market is assessed, and that demand for the processed product is determined. A surplus of fresh food is not sufficient reason for starting production.

Pineapples left to rot in time of glut

Many indigenous fruit and vegetable products such as fruit leathers, fruit pastes/jams, pickles, and dried chips are made in the home. These preserved products are usually stored for future use and are often not intended for sale.

In some countries, the advance of tourism and the growth of an urban middle class has produced a developing market for a wide range of western-style products including jams, jellies, and crystallized fruit sweets. Existing products are often made on an industrial scale, have sophisticated packaging, and are promoted by extensive marketing campaigns. Therefore, if a small-scale producer is aiming to compete effectively, the maintenance of high quality, in terms of both product and packaging, and keeping costs low is essential.

Nutritional significance

Fruits and vegetables provide an abundant and cheap source of vitamins, minerals, and fibre. Their importance in the diet is largely determined by culture, for example, a religion such as Hinduism demands that its followers are vegetarian and their diet therefore contains a high proportion of fruit and vegetables. Other communities, however, only serve vegetables as accompaniments to main meals, and fruits as appetizers and desserts.

It is preferable to consume fruit and vegetables when fresh, as the nutritional content is then usually at its highest. Some techniques, such as blanching, leach out many water-soluble vitamins into the surrounding liquid and if this liquid is not consumed, many nutrients are lost. Other methods such as sterilization expose the food to high temperatures which destroy some of the B vitamins. The table below illustrates the stability of nutrients, when exposed to certain processing or storage conditions.

Stability of vitamins under different conditions

Nutrient	Air	Light	Heat	Maximum cooking losses (per cent)
Vitamin A	U	U	S	40
Vitamin C	U	U	U	100
Biotin	S	S	U	60
Vitamin D	U	U	S	40
Vitamin K	S	U	S	5
Pyridoxine (Vitamin B6)	S	U	U	40
Riboflavin (Vitamin B2)	S	U	U	75
Thiamine (Vitamin B1)	U	U	U	55

S = stable (no important destruction)
U = unstable (significant destruction)

The importance of these losses depend on the actual nutritional status of an individual consumer (for example, the nutrients they are lacking, and the amount needing to be consumed).

Processing

Although there are many similarities between the processing of fruit and vegetables, it is important to realize the following differences.

3

Fruits are nearly all acidic and are commonly called 'high acid' foods. This acidity naturally controls the type of micro-organisms that are able to grow in fruit products. The spoilage micro-organisms that are likely to be found in such products are moulds and yeasts, which if consumed, rarely cause illness. Processing may be achieved by using preservatives such as sugar, salt and vinegar, and by drying, concentration or fermentation.

Vegetables are less acidic than fruits and for that reason are classified as 'low acid' products. A wide range of micro-organisms are able to grow in moist low-acid products, which may lead to spoilage and the possibility of food poisoning. To prevent this, vegetables can be processed by heating to destroy bacteria, or by pickling, salting, or drying to inhibit bacterial growth. Care is needed when processing low acid products, such as vegetables, to minimize the risk of transmitting food poisoning bacteria to consumers.

Jams, jellies, marmalades and fruit cheeses

Collectively known as preserves, these products are finding an increased importance in many countries, particularly in the more affluent urban areas. Fruit is most commonly used as the raw material, but some vegetables such as pumpkin can be used.

The principles of preservation involve heating to destroy enzymes and micro-organisms, combined with a high acidity and sugar content to prevent re-contamination. The mix of ingredients is quite complex, but basically involves the correct combination of acid, sugar, and the gelling compound 'pectin' (pectin is present naturally in plants, but may also be added in a commercially-produced form), to achieve the desired gel structure. The ingredients are then boiled together to evaporate water and achieve the correct sugar content.

Jams

This is a solid gel made from fruit pulp or juice, sugar, and pectin. It can be made from a single fruit or from a combination, but in either case the fruit content should be at least 40 per cent. In mixed-fruit jams, the first-named fruit should be at least 50 per cent of the total fruit added (based on European legislation). The total sugar content of jam should not be less than 68 per cent to prevent mould growth after opening the jar.

Jellies

These are crystal-clear jams, produced using filtered juice instead of fruit pulp.

Marmalades

These are produced mainly from clear citrus juices and have fine shreds of peel suspended in the gel. Commonly-used fruits include limes, grapefruits, lemons and oranges. Ginger may be used alone or in combination with the citrus fruit. The fruit content should not be less than 20 per cent citrus fruit, and the sugar content is similar to jam.

Fruit cheeses

These are highly boiled jam-like mixtures that have a final sugar level of 75–85 per cent and thus set in a solid block. They can be cut into bars or cubes, or further processed as ingredients in confectionery or baked goods.

The table below outlines the stages needed for the production of some fruit and vegetable products:

Production stages for some preserves

Process/product	Prepare raw material	Pulp/ extract	Sieve	Strain	Add other ingredients	Boil	Fill	Pack
Jam (whole-fruit)	*				*	*	*	*
Jam	*	*	*		*	*	*	*
Jelly	*	*		*	*	*	*	*
Marmalade	*	*		*	*	*	*	*
Fruit cheese	*	*	*			*	*	*

Equipment required

Processing stage	Equipment	Section reference
Prepare raw material	Fruit and vegetable cleaners	14.1
	Peeling machinery	51.0
	De-stoners	21.0
	Fruit and vegetable choppers	12.1
	Cutting, slicing and dicing equipment	17.0
Pulp/extract	Fruit presses	53.1
	Pulpers/juicers	55.0
	Juice centrifuges	07.3
Sieve/strain	Sieves	29.3
	Strainers	29.4
Add other ingredients	Weighing and measuring equipment	64.1 and 64.2
	pH meters	52.0
	Refractometer	56.0
Boil	Boiling pans or Steam jacketed pans	48.0
	Thermometer	63.0
	Heat source	36.0
Fill	Solid filling machine	28.2
Pack/seal	Sealing machine	47.1
	Capping machine	47.2

Cutting pineapple

Preparation of the raw material

The process begins with washing the incoming raw materials, and it is vital that the water used is potable. In some regions it may be difficult to obtain a supply of clean water and it may be necessary to purify it. This can be carried out by boiling the water and allowing it to settle, but this is a slow process and is costly in terms of fuelwood. It is possible to purchase water-filters or make your own. In all cases it is advisable that local expert advice is sought on the suitability of the water in any location before processing takes place.

Pulping/juice extraction

Juice can be extracted in a number of ways:

○ by steaming the fruit

○ by reaming the fruit (for citrus fruit)

○ by pressing

○ by pulping, using purpose-made pulpers, blenders, or a pestle and mortar.

Straining

The starting material for the production of jellies is a clear juice. To achieve this, the extracted juice must be strained using a muslin cloth bag. Additionally, sugar syrups should be strained in order to remove any unwanted material.

Addition of ingredients and process control

As in all processing, it is necessary to ensure that the correct amounts of ingredients are added, and that temperatures and other process conditions are standardized. This will ensure that the product has constant quality time after time. To standardize recipes successfully it is useful to have the following:

○ a pH meter or pH papers for checking the acidity level (the optimum range is 3.0–3.3)

○ a thermometer for temperature measurement

○ accurate scales for the measurement of small amounts of ingredients such as preservatives

○ a refractometer for accurately assessing sugar content. (Note: the boiling temperature can also be used as a less accurate measure of sugar content. The advantage is that a thermometer is cheaper than a refractometer.)

Boiling

Boiling can be carried out in a stainless steel or enamelled metal pot. If pans made from other materials are used there is the possibility that the fruit acids will react with the pan and cause 'off' flavours. For larger production it is best to use a steam jacketed pan.

There are two heating stages in the manufacture of jam. Initially, it is necessary to heat the fruit slowly in order to soften the flesh and extract the pectin. Once this is completed, it is vital to boil the mixture rapidly. This change in

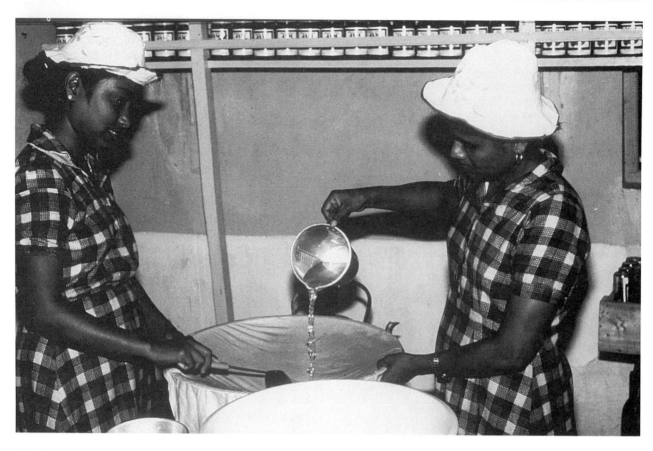

Filtering sugar syrup

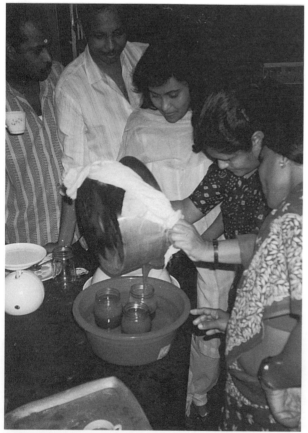

Filling jam into bottles

heat output is difficult to achieve without an easily-controllable heat source and the choice of heat source should be an important consideration before embarking on production.

During boiling care must be taken to avoid localized over-heating which is likely to lead to burning and colour change. Boiling is carried out until the desired sugar content is reached. There are various ways to test for this. It can be conveniently measured using a hand-held refractometer, or a sugar thermometer. These may be out of financial reach for a small producer, but if a product of consistent quality is to be produced, batch after batch, their use is highly recommended. Alternative checks, including placing a drop of the product in cold water to see if it sets, are less accurate, and require experience and skill to work effectively.

Filling and packaging

A good product packed in a dirty container will soon deteriorate, therefore it is essential that the containers used are thoroughly washed and sterilized. (For details refer to the Packaging chapter.)

The preserve should be hot-filled into suitable containers which are then sealed with a lid. The temperature of filling is important – too hot, and the steam will condense on the inside of the lid and drop down onto the surface of the preserve. This will dilute the sugar on the surface and allow mould growth. If the temperature is too low, the preserve will thicken and be difficult to pour and a partial vacuum will not form in the jar. Ideally the temperature should be 82–85°C.

Filling can be achieved using jugs and simple funnels. For higher production rates small hand-operated or semi-automatic piston-fillers are available. In all cases, the jars should be filled to the correct level, approximately 9/10ths full, to assist the formation of a vacuum as the product cools.

Finally, the jars are held upright while the gel is formed during cooling. This can be done by standing the jars on shelves, or, more quickly, using a low-cost water-cooler (see Packaging chapter for more details). A partial vacuum should form between the surface of the jam and the lid when the product cools. This can be seen by a slight depression in the lid. If a vacuum does not form, it means that the jar is leaking or the filling temperature was too low.

The packaging is likely to be one of the main costs involved in production. Ideally, glass jars should be used, with new metal lids. It is possible to use paper, polythene, or cloth tied with an elastic band or cotton, to cover the jars. The appearance of the product is, however, less professional, and there is a risk of contamination by insects. This is not recommended unless metal lids are impossible to obtain.

Increasingly, products are being packaged in pots with aluminium foil lids by large manufacturers. These packs are becoming popular with urban consumers as they are cheaper and more convenient than glass. However, they are difficult to obtain in many countries and are often expensive. Alternatives include plastic pouches/sachets. Technical advice should be sought if these packs are being considered.

Chutney, pickles, and sauces

These products are popular in some regions where they are used as accompaniments to meals. There are hundreds of varieties in existence and they can be made from a wide range of fruits and vegetables.

The basic principle of preservation for all of these products is the use of acetic acid (vinegar). Acetic acid preserves the product by making the environment acidic, and by so doing it inhibits the growth of spoilage and food-poisoning micro-organisms. Other ingredients such as salt and sugar add to the preservative effect.

Chutney
These are jam-like mixtures which have added vinegar and spices. The high sugar content exerts a preservative effect, and a high level of vinegar addition is not always needed. These products are hot-filled.

Pickles
Pickles can be either fermented or unfermented, sweet or sour, and can be made from either whole or chopped fruit.

Sauces
These are thick liquids made from pulped fruit and/or vegetables, with the addition of salt, sugar, and vinegar. They require pasteurization and are filled while hot.

The table (right) outlines typical processing stages for a representative range of products.

Product	Sweet pickle fermented	Sour pickle fermented	Unfermented pickle	Sauce
Select fruit/vegetable	*	*	*	*
Prepare fruit/vegetable	*	*	*	*
Add to brine solution and mix	* 5% brine + 1–2% sugar; leave for 1–2 weeks	* 5% brine; leave for 1–2 weeks		
Add salt			*	*
Mix with vinegar				*
Boil				*
Prepare vinegar mixture and boil	* 3% salt + 5% vinegar and sugar	* 3% salt + 5% vinegar	* 3% salt + 6% vinegar and sugar	
Pour hot vinegar mixture over the vegetables/fruit	*	*	*	
Pack (seal and label)	*	*	*	*
Pasteurize				*

Equipment required

Processing stage	Equipment	Section reference
Prepare raw material	Fruit and vegetable cleaners	14.1
	Peeling machinery	51.0
	De-stoners	21.0
	Fruit and vegetable choppers	12.1
	Cutting, slicing, and dicing equipment	17.0
Blanch	Steam blancher	01.0
	or boiling pan	48.0
	Heat source	36.0
Mix sugar and other ingredients	Weighing and measuring equipment	64.1 and 64.2
	Brine meter	64.6
Boil	Boiling pan	48.1
	Heat source	36.0
	Thermometer	63.0
Fill	Liquid filling machine	28.1
	Solid filling machine	28.2
Pack	Capping machine	47.2
	Sealing machine	47.1
Pasteurize	Water bath and large boiling pan	48.0
	Pasteurizer	50.0

Peeling apples

Preparation of the materials

Equipment needed for de-stoning and cutting are as indicated in the section for jams.

Addition of ingredients

To ensure that the product has a long shelf-life, it is necessary to balance the sugar concentration and acidity. To do this it is likely that the following pieces of equipment will be necessary:

○ pH meter

○ Brine meter

○ Refractometer.

It is possible to calculate a value known as the 'preservation index'. This is used to assess whether the product is safe from food spoilage and poisoning micro-organisms. The value can be calculated as follows:

$$\frac{\text{total acidity} \times 100}{(100 - \text{total solids})} = \text{not less than 3.6 per cent}$$

If you do not have access to basic laboratory equipment or are not sure how to carry out the calculation, it is best to take the sample to a food-testing laboratory and they will be able to tell you whether you need to adjust the recipe.

Pasteurization

Pickles, which have an adequate preservation index, do not need to be pasteurized. However, as an additional safety measure, it is common to boil the vinegar mixture, add it to the vegetables, and fill the product into the jars while it is still hot. In this way the hot mixture will form a partial vacuum in the jar and prevent re-contamination.

Sauces can be pasteurized before filling using a stainless steel pan or a steam jacketed pan, depending on the rate of production. Alternatively, pasteurization can take place after filling by placing the filled containers with the lids loosely on in a pan of boiling water and the water level around the shoulder of the jar. The time required for pasteurization will depend on the product, but most sauces are heated to between 80–95°C for five to ten minutes.

Filling and packaging

The same considerations for sterilizing and filling bottles apply as for jams. Glass jars are the most commonly-used packaging material. Pickles may also be packed in small quantities in polythene pouches. These simple pouches are sealed with a powered bar-sealer. To avoid seepage, it is suggested that a double pouch be used (i.e. an inner pouch made from food-grade polythene placed in an outer pouch made from cheaper polythene, and a label between the two).

Dried fruit and vegetables

Drying produce in the sun is simple and has the advantage of being a traditionally-understood technology with little or no fuel and equipment costs.

Drying removes water from the surface of the food by the combined effects of air flow, air temperature, and air humidity. The relationship between the three is important if drying is to be successful. When the moisture content is lowered below a certain level, micro-organisms cannot grow, and the produce is preserved.

In humid climates, dried products must be packaged well in order to prevent moisture uptake and protect against spoilage.

Air-dried products

These are the most common type of dried fruit and vegetables. Some products may be blanched or sulphured/sulphited to protect the natural colour and aid preservation. Dried fruit pulp is often named 'fruit leather'.

Dried and fried products

These are products which are partly dried, and then deep-fried, to produce a snack food. Examples include banana chips and bombay mix.

Osmotically dried fruits

These are fruits which are soaked in hot concentrated sugar syrups to extract some of the water prior to drying.

Production stages for dried fruits and vegetables

Process /product	Air-dried fruit	Air-dried vegetables	Fried/ dried product	Osmotically dried fruits
Prepare raw materials	*	*	*	*
Blanch		*		
Sulphuring/ sulphiting	*Some	*Some		
Prepare sugar syrup				*
Soak in syrup				*
Pulp	*Some (e.g. fruit leathers)			
Strain/filter	*Some (e.g. fruit leathers)			
Boil	*Some (e.g. fruit leathers)			*
Pour into thin sheets	*Some (e.g. fruit leathers)			
Dry	*	*	*	*
Deep-fry			*	
Pack	*	*	*	*

Equipment required

Processing stage	Equipment	Section reference
Prepare raw material	Fruit and vegetable cleaners	14.1
	De-stoners	21.0
	Peeling machinery	51.0
	Fruit and vegetable choppers	12.1
	Cutting, slicing and dicing machinery	17.0
Blanch	Steam blancher	01.0
	or boiling pan	48.0
Sulphur/ sulphite	Weighing and measuring equipment	64.1 and 64.2
	Sulphur cabinet	
Prepare sugar syrup	Weighing and measuring equipment	64.1 and 64.2
	Boiling pan	48.0
	Heat source	36.0
Soak in syrup	Boiling pan	48.0
	Food grade tank	03.1
Pulp	Pulper/juicer	55.0
Strain/filter	Muslin cloth	
	Stainless steel strainer/filter	29.0
Dry	Solar dryer	23.1
	Fuel-fired dryer	23.2
	Electric dryer	23.3
Deep-fry	Fryers	33.0
Pack	Sealing machinery	47.1

Processing notes

During drying, many fruits and vegetables experience some changes in colour. These can be lessened by carrying out some simple processing stages prior to drying (for example, blanching, sulphuring, and sulphiting).

Blanching is a short heating treatment in water or steam, and is often a necessary processing stage. It has many functions, but essentially it destroys enzymes which are responsible for causing browning, and reduces the total number of micro-organisms in the food.

For production on a small scale, the produce can either be wrapped inside a muslin cloth or in a wire basket, and immersed into boiling water. As the food is in direct contact with the water there is some loss of water-soluble vitamins. Steam blanching can be carried out by placing the produce in a strainer, which is then fitted over a pot of boiling water and covered with a lid to prevent the steam escaping. Steaming takes a few minutes longer than the water method but it has the advantage of losing fewer nutrients, as vitamins are not leached into the water. For larger production, a tray blancher can be purchased.

Sulphuring/sulphiting

With some dried products, the use of chemical preservatives will improve the colour and increase the shelf-life. The most commonly used preservative is sulphur dioxide. There are two methods: sulphuring and sulphiting. Sulphuring is more commonly used for fruits, and sulphiting for vegetables.

Sulphiting involves the use of sulphite salts, such as sodium sulphite or sodium metabisulphite. They may be either added to the blanching water or more commonly used by soaking the food in a solution of the salts.

The strength of a sodium metabisulphite solution is expressed in 'parts per million' (ppm) and the strength used will depend both on the final product required and the legal standards set in any particular country.

Sulphuring is achieved by burning sulphur in a sulphur cabinet. This can be made from locally available materials. The amount of sulphur used and the time of exposure depend on the commodity, its moisture content, and the levels permitted in the final product. The food is placed inside the cabinet and sufficient sulphur is placed in a container near the trays. For most vegetables 10–12g of sulphur (2–2½ level teaspoons) per kg of food is adequate. The sulphur is ignited and allowed to burn in the enclosed cabinet for 1–3 hours.

Drying techniques

The simplest method of drying is to lay the foods in the open air, either on mats, or on raised platforms. Although this is effective, there is limited control over the drying process which results in a variable product quality and a greater risk of contamination. To give more control over these aspects, solar dryers have been designed which protect the product from dirt and insects and increase the rate of drying.

Solar dryers fall into two categories – direct or indirect. In a direct dryer, the product is exposed to the sun's rays. This exposure results in vitamins being lost and a darkening in the colour of some foods. This colour change is desirable for products such as dates, but for lighter fruits, such as papaya and apricots, it is a problem.

Indirect dryers shelter the product from the sun. The heat from the sun is collected in a separate connected chamber and the heated air is passed over the food in an enclosed dryer.

Designs are also available for combined dryers. These are fitted with both a heater unit and a solar collecting chamber. When there is plenty of sunshine, the solar collector can be used, but the heater can also be used in poor weather conditions and at night.

There have been numerous designs for solar dryers, but most have met with a very poor response from rural producers. Most rural consumers are not willing to pay more for a slightly improved product quality and the investment in a dryer may not prove to be economically advantageous. In addition the relatively poor control over drying conditions compared to that for fuel-fired or electric dryers, means that they are largely unsuitable for high-value products such as spices, where an improvement in quality does generate higher income.

There are also a large number of designs of fuel-fired dryers. These have better control over drying conditions and therefore produce a higher quality product. They are able to operate at all times of the day and year, and in most cases produce a higher rate of drying. However, these benefits must be evaluated against higher capital and operating costs.

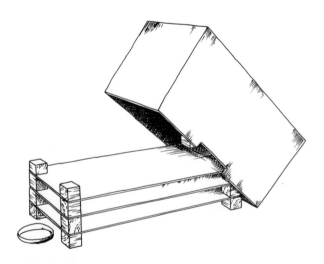

Sulphur cabinet

Packaging

Traditional packaging materials such as baskets, jute sacks, and wooden boxes have long been established for packaging dried foods such as fish and vegetables. They are for commodities which are transported in large quantities to a central marketing place and then sold loose. These packages can be used several times and are usually cheap.

Traditional packaging is only suitable provided the climate does not cause an increase in the moisture content of the food which will result in mould growth. If the climate is not suitable, dried foods should not be transported in this way. Boxes are used to prevent crushing of dried foods, and in humid climates, moisture-proof flexible films can be used (see Packaging chapter).

Some semi-moist foods such as osmotically dried fruits have special needs to prevent the reabsorption of water. Since dried fruit is a valuable product, it may be worth spending more on the package, such as a moisture-proof sealed bag. A wide range of flexible packaging materials is also available, but the use of many of these is limited due to high costs. Low-density polyethylene is a moderately good moisture barrier and cheaper than other films. It can be easily sealed using a powered bar-sealer.

Flexible materials may be used as the sole component of a package, but for most foods, a sturdy outer container is also needed to prevent crushing or to exclude light.

Suitability for small-scale production

It is technically feasible to make most fruit and vegetable products on a small scale using simple machinery, but it is likely that a group starting up in business will require substantial advice.

A common problem for small producer groups is the lack of market research. Such enterprises are often production-led, and products may be manufactured in order to use up a glut before a definite need or market for the product has been identified. Therefore, marketing will require special emphasis as this is often the most serious problem facing a new business. Rural production of value-added fruit products for urban or middle class markets has the added complication that the markets may be a long way from the producer group which may cause difficulties in negotiations and language problems, packaging supplies, and high distribution costs.

It is a common mistake to assume that poor-quality fruits and vegetables can be used to make high-quality goods. It is only possible to use rejected produce if it has been rejected for cosmetic reasons (e.g. the wrong size or slight blemishes).

For year-round production, it may be necessary to part-process raw materials into a form that can be stored in readiness for future production. Alternatively, a sequence of fruits or vegetables can be processed throughout the year in some regions. Both methods can help overcome the highly seasonal nature of fruit and vegetable crops. Despite this, in many cases processors will need a high working capital to buy the majority of raw materials in mid-season when prices are at their lowest.

A constraint in the production of preserves is that they require a large quantity of sugar. In many cases, refined white sugar has to be brought from urban centres, and may be expensive.

These points are not meant to discourage anyone from starting such a venture, but the problems should not be under-estimated, and it is best to seek advice first from a qualified technical source.

2. Cereal and pulse-based products

Staple foods are those which are eaten regularly as part of the daily diet and nearly always include cereals or pulses. Rice, for example, is widely consumed in Asia, whereas beans and maize are more popular in many African, Latin American and Caribbean countries.

As a result of importation and price subsidies in some countries, wheat has grown in popularity and the demand for wheat-based products such as bread and pasta has increased. Consequently some indigenous grains are not being used to their full potential.

Owing to a low moisture content, cereals and pulses are relatively stable during storage and processing is not so much for preservation but rather to change the eating quality and add variety to the diet.

Nutritional significance

Both cereals and pulses are nutritionally important since they usually provide the bulk of the diet. They are also a relatively cheap source of energy, protein, and important vitamins and minerals.

Processing

Some businesses are set up to clean and package wholegrains and pulses. These businesses can be successful as there is very little need for equipment and consequently the required capital investment is relatively low. However, with all business ventures, a clear demand for the product must be identified.

Wholegrains are normally processed before being used as an ingredient to produce a range of products. Processing techniques include puffing, flaking and milling. (This chapter does not cover production of drinks made from cereals. For details please refer to the Beverages chapter.)

Puffed products

Puffed grains and pulses are often used as ingredients in breakfast foods or as a snack food. During puffing, grains are exposed to a very high steam pressure, which when rapidly released causes the grains to burst open. The equipment necessary for puffing involves the use of high pressures and steam. In view of this the operator must take special care when handling the equipment, in order to prevent accidents. Puffed grains and pulses may be further processed by toasting, coating, or mixing with other ingredients.

Flaked products

Before flaking, grains and pulses are softened by being partially cooked in steam. They are then pressed or rolled into flakes which are subsequently dried. Such partially-cooked cereals may be used as quick-cooking or ready-to-eat foods.

Puffed rice

Flaking rice

Both flaked and puffed grains are eaten crisp and need to be packaged in moisture-proof containers. The moisture content must be below 7 per cent, otherwise the product becomes soggy and tough.

Milled products

Grains and pulses are milled to produce flours which are used to produce many types of food. Three main types of mill are available:

○ Plate mill

○ Roller mill

○ Hammer mill.

The choice of mill will depend on the raw material and also on the envisaged scale of production. Section 41.0 of this book illustrates a range of different mills, but for details of the milling process and in-depth guidelines relating to choice of machinery, it is suggested that reference be made to the 1992 edition of *Tools For Agriculture*.

Flours and flakes can be used as starting materials for the following types of product.

Products

Doughs and batters

Doughs and batters are made from a combination of flour, fat, and water or milk. By including other ingredients such as sugar, yeast, fruit, and nuts there is plenty of scope to create a variety of products.

The major difference between doughs and batters is that the moisture content of a dough is lower than that of a batter, and consequently doughs have a heavier consistency and can be moulded into many shapes. Both doughs and batters can be fermented to produce many products. For example fermented maize products such as 'Kenkey' in Ghana and 'Bagone' in Botswana are considered to be staple foodstuffs.

The production of batters and doughs involves mixing ingredients together to a smooth and uniform consistency. This can be done manually or with a powered mixer. The type of powered mixer required will depend upon the product being prepared. For example, a batter will need a balloon whisk-type mixer whereas a dough will require a hook-type mixer. There are general purpose mixers available for use on a small scale, which have attachments for both products.

There are various ways in which doughs and batters may be subsequently processed.

Extrusion

Extrusion involves forcing food through a hole (or die) to produce strands or other desired shapes. Once the food has been extruded it is likely to undergo a series of further processes such as frying, boiling, or drying.

Examples of extruded doughs include snack foods such as bombay mix and a wide variety of pasta products and noodles. The latter are increasingly becoming part of the daily diet for many people around the world since they are quick and easy to cook, offer convenience, and require less fuel for cooking. A further advantage of these products is that if dried, they can be stored for a long time.

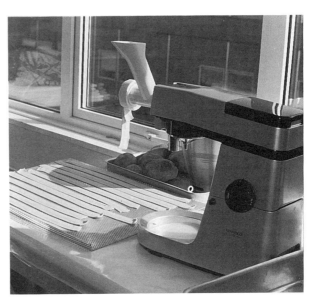

Pasta being extruded

The table below outlines stages in production for rice noodles. Similar processes are used for pasta and snack foods using appropriate ingredients:

Production stages for rice noodles

Ingredients	Processing stage	Equipment	Section reference
Rice flour, water	Mix ingredients to a dough	Powered mixer (optional)	43.2
	Roll into thin sheets of dough	Rolling equipment	59.0
	Cut into strands	Cutting equipment	17.1
	Extrude	Extruder	27.0
	Steam	Steam blancher	01.0
	Dry	Solar dryer Fuel-fired dryer Electric dryer	23.1 23.2 23.3
	Pack	Polythene sealing machine	47.1

Noodles can be made from a range of flours including rice, wheat, maize, and potato. The dough can be processed in one of two ways, either rolled out into thin sheets of dough and cut into strands, or extruded.

Cereal and pulse-based products

Preparation of noodles

The strands are then steamed, and may either be eaten fresh, or processed further by drying. Drying can be achieved using either a solar or a fuel-fired dryer. It is possible to dry the noodles in the sun, but the quality of the finished product is likely to be lower. Packaging requirements for dried noodles include a moisture-proof package (e.g. polythene) and an outer carton/box to prevent crushing.

Baking

A wide assortment of baked products can be produced, including cakes, biscuits, and leavened bread. For details of processing and equipment requirements, please refer to the Baked Goods chapter.

A further group of products include a range of unleavened breads which are baked using a hotplate, often known as a griddle. Doughs are usually rolled to shape, (e.g. chapattis, roti, and tortilla), whereas batters are usually dropped onto the griddle using a spoon (e.g. pancakes and 'Kisera', a Sudanese product made from sorghum).

The table opposite outlines the stages in the production of tortilla, a bread widely consumed in Latin America.

Cooking chapattis

Frying

Frying involves cooking food in hot oil. Doughnuts are an example of a fried dough which is consumed as a snack food.

It is important to make sure that the temperature of the oil is correct. If the temperature is too low then the level of

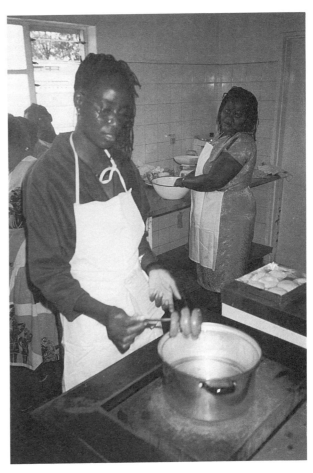

Frying doughnuts

Production stages for tortilla

Ingredients	Processing stage	Equipment	Section reference
Maize, lime*	Boil grains in water and lime (until the grains can be peeled by hand)	Boiling pan	48.0
	Wash the maize	Clean potable water	
	Mill grains with the addition of water	Plate mill	41.1
	Mix to a dough with water	Powered mixer (optional)	43.2
	Shape	Rolling equipment	59.0
	Fry on a hotplate	Hotplate	36.0
	Consume immediately or pack	Heat sealing machine	47.1

*The lime involved is not the fruit but the chemical.

Production stages for doughnuts

Ingredients	Processing stage	Equipment	Section reference
Flour, fat, yeast, sugar	Mix dough	Mixer	43.2
	Shape		
Jam	Add jam if necessary		
	Place on oiled tray and prove	Proving cabinet	
Oil	Fry	Deep-fat fryer	33.0
	Drain		
Sugar	Roll in sugar		
	Pack	Polypropylene or polythene bags if necessary Sealing machine	47.1

Preparing maize porridge

fat absorption in the doughnut will be too high, leading to a greasy product. If, on the other hand, the temperature is too high, then it will become necessary to remove the doughnuts before they are cooked, in order to prevent over-browning.

The shelf-life of fried foods is mostly determined by the moisture content after frying. For example, doughnuts have a relatively short shelf-life owing to moisture and oil migration during storage. Packaging is therefore not necessary, except to keep the product clean.

Porridge-type products

Flours and flakes are often cooked with water to produce a porridge. These porridges are prepared in the home as a main meal, and their consumption is prominent in many parts of Africa (e.g. foodstuffs such as 'banku' and 'ugali' made from maize and consumed in Western and Eastern Africa respectively).

Weaning foods are a combination of cereals and pulses prepared as a mixture of flours and flakes. These foods are designed to be made into a porridge by adding water or milk. Many weaning foods are made by multinational manufacturers on a large scale. However, it is possible to produce a weaning food on a small scale by:

○ preparing the correct combination of raw materials

○ roasting the mixture of cereals and pulses

○ milling using a hammer mill

○ packing as a dry mix.

Such mixtures can then be reconstituted in the home and mixed to a porridge-like consistency with water. It is vital that nutritional advice is sought to ensure that the mixture is nutritionally balanced, containing sufficient energy, protein, and other nutrients which will maintain the health and growth of the child. Such foods are vital since it is in this transition period from the breast to the family pot that the child is most likely to suffer from protein-energy malnutrition and other nutritional deficiency diseases.

The suitability for small-scale production

It is possible to make many cereal- and pulse-based products on a small scale using domestic scale equipment. Some products such as pasta require special pieces of equipment, but the majority are not overly expensive and are affordable by many small-scale producers.

The technical knowledge required for many cereal and pulse-based products is quite low, but consistent quality and a degree of flair and skill is needed to produce a marketable product.

3. Baked goods

Baked goods are widely available in many countries. Generally they fall into the main categories of breads, biscuits, cakes, and pastries, and are consumed by people from most income groups.

Selling baked goods from a bicycle

Baked goods also make up a proportion of street foods. In this form, they offer convenience and provide a cheap source of energy, which is important for those who travel to work and rely on such foods for their first meal of the day.

Grains and flours have a relatively long shelf-life. The purpose of baking is therefore not for preservation, but to change the eating quality of staple foodstuffs and to add variety to the diet. The growth of food-poisoning organisms is less likely to occur in baked goods than in other low-acid foods such as meat, milk, and fish. However, it is still necessary to check the quality of raw materials and hygienic conditions during processing.

Baked goods such as breads and pastries have a shelf-life in the range of 2–5 days, whereas others, such as biscuits and some types of cake, have a shelf-life of several months, when correctly packaged.

Nutritional significance

The main ingredients in baked goods are flours from cereals such as wheat, maize, and sorghum. In general, all flours contain valuable amounts of energy, protein, iron and vitamins, but the degree of milling will influence the final nutritional content.

Traditional milling produces a flour which contains all of the crushed grain. Although these flours often have a coarse texture and an off-white colour, they contain many B vitamins, minerals, and also fibre. The desire for white flour however, has led to a milling process which removes the bran and, as a result, many nutrients are lost.

Many baked products incorporate high levels of fat, sugar, and sometimes fruit or nuts, and this will increase the energy content of the products.

Production

Baking is a process which uses heated air to alter the eating quality of foods. A secondary purpose is preservation, by destroying micro-organisms and reducing the moisture content at the surface of the food.

In this chapter, 'baked goods' refers to breads and flour confectionery, but it is also possible to apply the process to meat, fruits, and vegetables. The process in these cases is often referred to as 'roasting'.

Baked goods are produced from either doughs or batters, which are a mixture of flour and water made by mixing, beating, kneading, or folding. The process will depend on the product being made and the ingredients used.

The tables overleaf outline the types of process and the equipment required to make a representative range of products.

Baked goods

Processing stages

Process/product	Mix	Ferment	Form	Prove	Knock back	Bake	Pack
Leavened bread	*	*	*	*	*	*	*
Buns	*	*	*	*	*	*	*
Biscuits	*		*			*	*
Pastries	*		*			*	*
Cake	*		*			*	*
Unleavened bread	*		*			*	*

Equipment required

Processing stage	Equipment	Section reference
Mix	Liquid mixer Solid mixer	43.1 43.2
Ferment	Proofing cabinet	
Form	*Bread:* Dough moulder *Biscuits:* Rolling machinery Depositor Dough cutter *Pastry:* Rolling machinery Pastry mould *Cakes:* Baking tins	 45.2 59.0 20.0 45.2 59.0 45.2 45.3
Prove	Same as for the fermenting stage	
Knock back		
Bake	Oven	46.0
Pack	Polyethylene or cellophane wrapping/sealing machines	47.1

Processing

Raw materials

Wheat flour contains a substance called gluten. This is responsible for providing the internal structure of products such as bread and some biscuits. Depending on the variety and the climate in which it is grown, wheat will contain different quantities of gluten. For example, flours which contain low levels are known as 'soft' or 'weak' flours, and those with high levels are known as 'strong' flours.

The type of wheat flour used will depend on the product being made. Bread and biscuits require a strong flour, whereas cake flours are usually medium to weak in strength.

Flours from cassava and rice may be used as a substitute for wheat, but as they do not contain gluten, a maximum substitution rate of 25 per cent is possible before differences in taste and texture become very noticeable. Additionally, there are many other products that are made completely from flours such as maize or rice.

Mixing dough

Mixing

All doughs and batters are mixed to achieve a smooth consistency and to ensure an even distribution of ingredients. This is usually done using simple hand-operated tools, but if a large quantity is to be produced it may be more convenient to use a powered mixer.

Mixers are classified according to the type of product that they are used for, i.e. either solids or liquids. A common solids mixer operates with a planetary action and can be fitted with a hook for pastry or dough, or a 'K-beater' for dry powders. Liquid mixers include a balloon-whisk attachment for batters or propeller-type mixers for thin liquids such as sugar solutions.

Fermentation

For some products it is not necessary for the dough to rise, or for the final texture to be open and porous. Such examples are unleavened breads such as roti, chapatti, and tacos, which as the group-name suggests do not require fermentation. Other products, such as scones or soda bread, rely on gas produced by added sodium bicarbonate to raise the dough and produce the desired crumb texture. In cakes, air is incorporated into the batter during mixing to give an open structure to the crumb after baking.

Only bread and buns are fermented to produce the open porous crumb. In this case added yeast ferments (breaks down) the sugars in the flour (and/or added sugar) to produce a gas, called carbon dioxide, which raises the dough. No special equipment is required to bring about fermentation. The dough is covered with a damp cloth or polythene and left to rest at a temperature of 32–35°C until it becomes fully inflated with carbon dioxide. To achieve this, the dough may be left near to the oven where it is warm, or in a temperature-controlled cabinet known as a proving cabinet.

Proving is the name given to a process of secondary fermentation which is applied to bread and bun dough. It is necessary because as the dough is being shaped some of the carbon dioxide is lost and the dough structure partially collapses. The aim of proving is to ferment the dough a little more and regain the carbon dioxide levels and therefore the open structure.

Kneading/forming/moulding

Kneading is a process of stretching and folding dough. In doing so the gluten fibres are stretched and the consistency of the dough becomes smooth. Kneading is most usually done by hand, but if a large quantity is being produced it can be a tiring task and a powered kneader may be preferred.

After kneading, the dough has to be formed into the desired shape and size. The method for doing this will depend upon the type of product, for example, bread dough can either be cut, folded, or plaited into shape. For production above the household level, a machine known as a dough divider can be used to cut the dough into accurately-weighed pieces, out of which the individual loaves or buns will be produced.

Biscuit doughs are rolled out and cut using either a knife or a shaped biscuit cutter. Pastry dough can be handled in a similar way. In the production of pies, a pastry mould is used to retain the shape of the dough during baking.

Baking

During baking, food is heated by the hot air, and also by the oven floor and trays. Moisture at the surface of the food is evaporated by the hot air, and this leads to a dry crust in products such as bread and many biscuits. If a glazed product is desired then it will be necessary to either inject steam, or place a pan of water in the oven. Oven temperatures for baking depend on the type, size, and shape of the product, and the nature of the ingredients.

The table outlines some typical combinations:

Moulding dough

Oven temperatures for baking

Product	Degrees F	Degrees C	Time (minutes)
Cake	350	176	45–60
Pastry	450	232	15–20
Biscuits	425–450	218–232	10–15
Bread	400	204	30–40

A range of wood, charcoal, gas, or electric ovens is available. The choice will depend on the cheapest and most widely available fuel in a particular area.

Packaging and storage

In general, baked goods have a short shelf-life, but for products such as biscuits, this can be extended from a few days to several weeks, or months, if packaged correctly.

Many biscuits are characterized by their crispness, and as a result need a low moisture content. In order to prevent the biscuit from becoming soggy, it is important that the packaging material prevents the uptake of moisture from the surrounding air. In addition, biscuits that contain fat should be protected from light, air, and heat to prevent development

Putting bread into the oven

of 'off' flavours due to rancidity. Plastic films, glass jars, or metal tins are all acceptable packaging materials.

As bread is usually eaten within one or two days after purchase, the main purpose of packaging is to keep the bread clean. Simple paper or polythene wrapping is often used. If polythene is used, the bread should be packed after it has cooled to prevent condensation of water inside the pack and resulting wet spots which cause mould growth.

The packaging and storage conditions for cakes depend on their moisture content and the relative humidity of the surrounding air. Each cake must be judged by its composition and the intended shelf-life. Light cakes, such as those made from flour, sugar, and egg, have a shelf-life of only a few days if not packaged. Fruit-cakes however have a more dense structure and a longer shelf-life. This can be extended to several years by coating the cake in marzipan (ground almond paste) and icing sugar, both of which act as a moisture barrier.

Suitability for small-scale production

Baked goods provide plenty of scope for the producer to use locally available ingredients to create a variety of value-added products. In general, it is more profitable to produce buns, biscuits, and cakes than bread. However, it is very difficult to predict the profitability for such goods in specific areas, as prices for wheat flour vary from country to country and are heavily affected by price and import subsidies. Packaging requirements are often minimal as many of the products are for immediate consumption. This therefore reduces some of the problems for producers in terms of availability and access to packaging materials.

4. Snack foods

Snack foods are foods which can be eaten in place of, or in between, meals. They are convenient because they are quick and easy to eat. The term 'snack food' does not only apply to some of the newer products such as potato crisps, but it also includes many traditional food items.

Most snack foods are intended for immediate consumption and have a shelf-life of only 1–2 days. Such food products may be sold loose without packaging, or in small polythene or paper packages which contain a portion for sale. If required, the shelf-life may be extended considerably through the use of adequate packaging.

In most countries people from nearly all income groups consume snack food products. For example, a recent study in Indonesia revealed that the largest consumer groups in the urban areas were workers and schoolchildren.

Snack foods add variety to the diet which partially explains their popularity. They may also play a cultural role on special occasions or when offered to visitors.

Nutritional significance

Snack foods frequently receive criticism due to their high levels of salt, sugar, and fat. They are seen to be nutritionally damaging when eaten regularly in place of a traditional food.

Snack foods, however, can be very nutritious when made from fruits, pulses, or cereals. It should also be pointed out that the consumption of snack foods does not necessarily lead to health problems such as obesity, but the cause is rather an unbalanced diet with excess fat, sugar, and salt. Therefore, if these food products are part of a wider diet, they can be an important source of fats and energy, particularly for the poorer sectors of society whose diet may be lacking in these nutrients.

Often, it is cheaper to purchase snack foods than it is to make a meal at home. This is likely to be one of the reasons why poor people are relying more and more on such

Selling packaged snack foods

products. Along with the convenience factor, this may, for example, explain the increase in sales of sweet buns to workers in Bangladesh.

Thus, in relation to the accusation that snack foods are making the poor malnourished, it may also be that poverty is forcing poor people to rely on snack foods and thus malnutrition is really a symptom of poverty.

Selling loose snack foods

Processing

Snack foods are made from a wide range of raw materials and the preparation differs from product to product. Frying, however, is the main process by which many are made and this is considered in detail here. (It should also be noted that concentrated milk solids or fruit pulps are also used for snacks in some countries.)

Principles of frying

Frying alters the eating quality of food. It also provides a preservative effect as the heat treatment destroys micro-organisms and enzymes, and there is a reduction in moisture at the surface of the food.

Choice of oil

Most oils used for frying are of vegetable origin, but there is no reason why animal fats cannot be used. The oil used has a great impact on the taste, texture, and keeping-quality of the final product.

Fats and oils are subject to a type of deterioration known as rancidity. This produces disagreeable odours and flavours and makes the fried foods unpalatable. Some oils are more prone to rancidity than others, and this is important when considering which oil to use. In many countries, however, there is only one type of oil widely available at the lowest cost, and processors will use this, despite rancidity problems, if it gives a flavour that is acceptable.

Raw material preparation

Cutting

Cutting into slices or cubes is often the only preparation needed for many snack foods. An ordinary kitchen knife can be used, or one of the many gadgets on the market (both manual and powered), make this stage easier and faster.

Slicing potatoes

Forming

Some snack foods are prepared by forming a dough and then shaping it into pieces. These doughs are usually cereal- or pulse-based and may be mixed manually or by using a powered mixer. Shaping can take place by either rolling out and cutting or by extruding the dough into strands before frying. Small hand-held extruders are widely available in some countries and can be fabricated locally from materials such as wood or metal. They may be fitted with dies of

Extruding dough

different shapes and sizes to add variety to the products made.

Some doughs can be prepared from milk solids. This involves evaporating milk until it forms a thick granular mass. This is then mixed with flour and sugar to produce a dough which is subsequently shaped into balls and fried. Examples are 'Mhisti' in Bangladesh and 'Rasangolla' in India, both of which are served floating in sugar syrup.

Frying

The amount of oil required for frying will vary according to whether the product is to be shallow-fried or deep-fried.

The temperature to which the oil is heated is not limited by a boiling point as with water. Heated oil does however reach a stage at which it breaks down to fumes, which is known as the smoke-point. It is important that oils do not reach the smoke-point when used for frying as this will cause the oil to deteriorate more rapidly and increase the danger of it catching fire. Suitable temperatures for frying are between 180 and 200 °C.

Frying can take place using a simple pan heated by an open fire or other heat source. Alternatively, for deep-frying, an electrically powered fryer, fitted with a thermostatic control, gives more control over heating for larger quantities of food.

Draining

Fried products need to be drained adequately in order to remove any excess oil. If this is not achieved the excess oil can make the product soggy, which is particularly important if snack foods are characterized by their crispness. In addition, poorly-drained products are likely to leave a film of oil on the inside of polythene packs. Not only does this look unappetizing, but it will also promote more rapid rancidity.

Leaving the product to drain is rarely sufficient to remove excess oil. Fans may help remove more oil or it can be absorbed into paper. In general, higher temperatures during frying cause less oil to be retained on the product. New oil also sticks to the product less than old oil.

Packaging and storage

Snack foods which are sold for immediate consumption have no need for packaging. For a longer shelf-life the package should provide a barrier to moisture to avoid the product losing its crispness, and to oil, air, and light. As many fried foods are also fragile, the pack should help prevent crushing.

For long-term storage, paper is of no use for this type of product. Polythene, while being a good barrier to moisture and oil seepage, is not a good barrier to light or air. Polypropylene is a better alternative although it is often more expensive.

The following tables outline the production and equipment needed for a selected range of snack foods:

Muchorai

Muchorai is a fried, savoury snack made from green gram and rice flour. It consists of strands extruded into a ring and has a deep yellow colour.

Muchorai

Snack foods

Ingredients	Processing stage	Equipment	Section reference
Rice flour, green gram flour, salt	Weigh and measure ingredients	Weighing and measuring equipment	64.1 and 64.2
	Mix dry ingredients	mixer (optional)	43.2
Fat, water	Add fat and water		
	Mix	Dough mixer (optional)	43.2
	Extrude	Extruder	27.2
Oil	Deep-fry	Frying pan or deep-fat fryer	33.0
		Heat source	36.0
	Drain	Absorbent paper or fan (both optional)	
	Pack using polythene or polypropylene packs for longer shelf-life	Sealing equipment	47.1

Gulab jamun

Gulab jamun is a milk-based sweet made throughout Asia. It is round in shape and has a deep brown, slightly crisp, outer surface, and is soft and porous inside.

Ingredients	Processing stage	Equipment	Section reference
Milk	Weigh and measure ingredients	Weighing and measuring equipment	64.1 and 64.2
	Evaporate the milk	Boiling pan	48.0
Baking powder, flour, sugar, water	Add dry ingredients		
	Mix	Mixer (optional)	43.2
	Add water	Mixer with dough hook (optional)	43.2
	Shape		
Oil	Shallow fry	Frying pan heat source	36.0
	Drain	Fan or absorbent paper (both optional)	
Sugar, water	Transfer to sugar syrup		

Gulab jamun

Potato crisps

These are potato slices which have been deep-fried. They may be flavoured and sold packaged in a polypropylene bag.

Ingredients	Processing stage	Equipment	Section reference
Potatoes	Wash and peel potatoes	Vegetable cleaners	14.1
		Peeling equipment	51.0
	Slice potatoes into thin slices	Slicing equipment	17.2
Oil	Deep-fry potato slices	Frying pan or deep-fat fryer	33.0
	Drain	Fan or absorbent paper (both optional)	
Flavourings e.g. salt	Flavour		
	Pack into polypropylene bags	Sealing equipment	47.1

Frying jalabees (Bangladeshi sweets)

Suitability for small-scale production

Snack foods are well suited to small-scale production for the following reasons:

○ There is a relatively small investment in equipment. Typical kitchen tools can be used to make most products.

○ Most snack foods are made from pulses, cereals, milk, fruits, and vegetables. These are usually readily available in the locality.

○ The technology is relatively simple and usually well known.

○ Adding value to basic raw materials by processing them into snack foods is often a profitable form of employment for small-scale producers.

The range of shapes, colours, flavours, and sizes of snack foods is almost infinite. This allows producers with flair and imagination to develop their own individual products, and develop their businesses.

5. Honey, syrups and treacle

Before the introduction of refined white sugar, products such as honey, syrups, and treacle were used as the main sweetening agents. Today they are commonly used to sweeten beverages such as tea and coffee and as ingredients for commodities such as sugar confectionery.

This group of products has a fairly long shelf-life if processed and packaged correctly. With a high sugar level of approximately 78–84 per cent, microbial activity is restricted and the product is stable for many months.

Man tapping a kitul palm

Honey

Honey is a sweet viscous syrup produced by honeybees. Bees deposit nectar into honeycombs and seal them with beeswax to preserve the honey. Honey is made up of a solution of sugars (glucose and fructose) and minerals in water, and is twice as sweet as sugar (sucrose). The composition varies according to many factors including:

○ The time of the year that the nectar is collected.

○ The weather at the time of collection.

○ The source of flower nectar.

Syrups and treacle

Both syrup and treacle are viscous liquids. They can be extracted from many different plant sources, including coconut and kitul palms, maple, and sugar-cane. Processing involves extracting the juice (sap) from the plant, and heating it to remove water. By doing this, the sugar content is increased, along with the stability of the product.

Nutritional significance

These products are rich in sugar and therefore provide energy. They also contain vitamins and minerals, but in some cases levels are too low to have any major nutritional significance. As these products are not highly refined they contain many other micro-nutrients which are beneficial to the diet. Honey, for example, is said to contain naturally-occurring antibiotics which may have a positive effect on the health of the body.

Processing

The table below shows the stages involved in the processing of honey, syrups, and treacle.

Processing stages for honey, syrups, and treacle

	Extraction	Filtration	Honey conditioning	Boiling	Packaging
Liquid honey	*	*	*		*
Tree syrups	*	*		*	*
Treacle/molasses	*	*		*	*

Equipment required

Processing stage	Equipment	Section reference
Extraction	Collection vessel/pan	
	Fruit juice extractors	53.1
	Honey centrifuge	07.2
Filtration	Filter cloth	
	Stainless steel mesh	
	Strainers	29.3 and 29.4
Honey conditioning	Fan	
Heating/ boiling	Boiling pan	48.0
	Heat source	36.0
	Thermometer	63.0
Packaging	Containers	62.0
	Filling machines	28.2
	Capping equipment	47.2

Extracting the sap

Sap is removed from plant sources such as coconut and kitul palm through a process known as 'tapping'. This involves cutting the flower, and allowing the sap to drain into a collecting bowl. It is important that the sap is processed quickly after collection, in order to prevent the onset of fermentation (this will reduce the level of sugar in the sap).

Fermentation is likely to be a problem for some producers as the trees are often some distance from one another which may cause the sap to be left for too long prior to being processed. Additionally, collection bowls may be inadequately cleaned, causing a build-up of yeast which will again cause fermentation. Preservatives such as sodium benzoate and sodium metabisulphite can prevent the onset of fermentation, but with proper hygiene their use is unnecessary.

Other plant material such as sugar-cane may be crushed with a roller mill to extract the juice.

Extracting honey

The initial processing stage involves removing the wax from the honeycombs. This is done with a special uncapping knife, which may be heated by steam or simply dipped in hot water to melt the wax.

There are several methods by which the honey can be extracted from the honeycombs. At a basic level, the honeycombs may be crushed by hand. This, however, is not recommended as it is difficult to remove the comb fragments and the resulting honey has a lower quality. Alternatively, honeycombs may be piled on top of a plastic sieve with a stainless container underneath for approximately three days, by which time the honey will have collected in the container. Although this method is effective, it takes time and may allow the product to be contaminated by insects.

Many producers use either a manual or a powered honey centrifuge. After removing the wax, the combs are placed in a frame which is placed in the centrifuge. The spinning action throws the honey from the combs and it is collected inside the centrifuge.

Filtration

Honey, syrups, and treacle often contain foreign matter, such as pollen seeds, bee hairs, and vegetable matter. They can be filtered out using a muslin cloth or a fine stainless steel mesh. These viscous liquids are easier to filter if they are warmed to 26–37°C so that they flow more easily.

Honey conditioning

Some honey producers carry out a processing stage known as conditioning to remove excess moisture and to concentrate the sugar to 82 per cent. This may be achieved by gently heating the honey in an open boiling pan to a temperature of no more than 50°C. There are many drawbacks to this method, as the product will require constant stirring to avoid localized over-heating, which may result in darkening of the product. There is little temperature control, and 'off' flavours may develop if the temperature gets too high.

An alternative technique involves blowing air at 25–40°C over the surface of the honey with a fan for several hours until the optimum moisture content of 18 per cent is reached.

The sugar content of the honey can be checked using a refractometer. Most refractometers calculate sugar concentration on the basis of sucrose. Honey, however, consists of fructose and glucose and therefore necessitates the use of a specifically-designed honey refractometer.

Heat treatment

Syrups and treacle are often heated to reduce the water content and as a result increase the sugar content. The period of heating should be carefully controlled to obtain the correct sugar level. If the sugar content is too low (i.e. below 75 per cent), fermentation can occur. Heating also caramelizes the sugars, giving the product a darker colour.

Heating may be achieved using an boiling pan, but due to the delicate flavours and colours of these products it is preferable to use a steam jacketed pan. Although the latter are more expensive, they are better able to maintain the quality of the product by avoiding localized over-heating.

Refractometers can be used to measure the sugar concentration and therefore ensure that the concentration process has been adequately achieved.

Packaging

All of the aforementioned products can be packaged in traditional materials such as clay pots, sold directly out of bulk containers, or bottled in glass jars/bottles. If bottled, it is important that the bottle is securely sealed to prevent leakage and prevent contamination by insects.

Suitability for small-scale production

Small-scale production requires little in terms of equipment, particularly if the product is sold in traditional packaging materials.

Honey, syrups and treacle

To produce a product with a consistently high quality, it is suggested that the producer use a refractometer. This piece of equipment may increase the capital investment but it is necessary to ensure high product quality.

Bottled treacle on sale

6. Sugar confectionery

Sugar confectionery refers to a large range of food items, commonly known as sweets. Boiled sweets, toffees, marshmallows, and fondant are all examples.

Sweets are a non-essential commodity, but are consumed by people from most income groups. The variety of products is enormous, ranging from cheap, individually-wrapped sweets, to those presented in boxes with sophisticated packaging.

Nutritional significance

The main ingredient used in the production of sweets is sugar (sucrose). There is a danger that if sweets are consumed in excess over a prolonged period of time they may contribute to obesity. Unless good dental care is practised, over-consumption can also lead to tooth decay.

Principles of sugar confectionery production

By varying the ingredients used, the temperature of boiling, and the method of shaping, it is possible to make a wide variety of products. In all cases, however, the principle of production remains the same and is outlined below:

○ balance the recipe

○ prepare the ingredients

○ mix together the ingredients

○ boil the mixture until the desired temperature has been reached

○ cool

○ shape

○ pack.

Many factors affect the production and storage of sweets:

A range of sweets for sale

○ the degree of sucrose inversion (see below)

○ the time and temperature of boiling

○ the residual moisture content in the confectionery

○ the addition of other ingredients.

Degree of inversion

Sweets containing high concentrations of sugar (sucrose) may crystallize either during manufacture or on storage (commonly referred to as graining). Although this may be desirable for certain products (such as fondant and fudge), in most other cases it is seen as a quality defect.

When a sugar solution is heated, a certain percentage of sucrose breaks down to form 'invert sugar'. This invert sugar inhibits sucrose crystallization and increases the overall concentration of sugars in the mixture. This natural process of inversion, however, makes it difficult to accurately assess the degree of invert sugar that will be produced.

As a way of controlling the amount of inversion, certain ingredients, such as cream of tartar or citric acid, may be used. Such ingredients accelerate the breakdown of sucrose into invert sugar, and thereby increase the overall percentage of invert sugar in the solution. A more accurate method of ensuring the correct balance of invert sugar is to add glucose syrup, as this will directly increase the proportion of invert sugar in the mixture.

The amount of invert sugar in the sweet must be controlled, as too much may make the sweet prone to take up water from the air and become sticky. Too little will be insufficient to prevent crystallization of the sucrose. About 10–15 per cent of invert sugar is the amount required to give a non-crystalline product.

Time and temperature of boiling

The temperature of boiling is very important, as it directly affects the final sugar concentration and moisture content of the sweet. For a fixed concentration of sugar, a mixture will boil at the same temperature at the same altitude above sea-level, and therefore each type of sweet has a different heating temperature (see chart below).

Boiling point of sucrose solutions

Sucrose concentration (per cent)	Degrees C Boiling point *	Degrees F Boiling point *
40	101.4	214.5
50	102	215.5
60	103	217.5
70	105.5	222
75	108	227
80	111	232
85	116	241
90	122	252
95	130	266

*at sea level.

Variations in boiling temperature can make a difference between a sticky, cloudy sweet or a dry, clear sweet. An accurate way of measuring the temperature is to use a sugar thermometer. Other tests can be used to assess the temperature (for example, toffee temperatures can be estimated by removing a sample, cooling it in water, and examining it when cold). The temperatures are known by distinctive names such as 'soft ball', 'hard ball' etc., all of which refer to the consistency of the cold toffee.

Type of sweet	Temperature range for boiling (Degrees C)
Fondants	116–121
Fudge	116
Caramels and regular toffee	118–132
Hard toffee (e.g. butterscotch)	146–154
Hard-boiled sweets	149–166

Moisture content

The water left in the sweet will influence its storage behaviour and determine whether the product will dry out, or pick up, moisture.

For sweets which contain more than 4 per cent moisture, it is likely that sucrose will crystallize on storage. The surface of the sweet will absorb water, the sucrose solution will subsequently weaken, and crystallization will occur at the surface – later spreading throughout the sweet.

Added ingredients

The addition of certain ingredients can affect the temperature of boiling. For example, if liquid milk is used in the production of toffees, the moisture content of the mixture immediately increases, and will therefore require a longer boiling time in order to reach the desired moisture content.

Added ingredients also have an effect on the shelf-life of the sweet. Toffees, caramels, and fudges, which contain milk-solids and fat, have a higher viscosity, which controls crystallization. On the other hand, the use of fats may make the sweet prone to rancidity, and consequently the shelf-life will be shortened.

Types of sweets

Fondants and creams

Fondant is made by boiling a sugar solution with the optional addition of glucose syrup. The mixture is boiled to a temperature in the range of 116–121°C, cooled, and then beaten in order to control the crystallization process and reduce the size of the crystals.

Creams are fondants which have been diluted with a weak sugar solution or water. These products are not very stable due to their high water content, and therefore have a shorter shelf-life than many other sugar confectionery products.

Both fondants and creams are commonly used as soft centres for chocolates and other sweets.

Gelatin sweets

These sweets include gums, jellies, pastilles, and marshmallows. They are distinct from other sweets as they have a rather spongy texture which is set by gelatin.

Toffee and caramels

These are made from sugar solutions with the addition of ingredients such as milk-solids and fats. Toffees have a lower moisture content than caramels and consequently have a harder texture. As the product does not need to be clear, it is possible to use unrefined sugar such as jaggery or gur, instead of white granular sugar.

Hard-boiled sweets

These are made from a concentrated solution of sugar which has been heated and then cooled to form a solid mass containing less than 2 per cent moisture. Within this group of products there is a wide scope to create many different colours, flavours and shapes through the use of added flavourings and colourings.

The table below outlines the processing stages for a selected range of confectionery items.

Processing

Boiling

There are three main ways by which to boil the sugar solution:

○ a simple open boiling pan

○ a steam jacketed pan

○ a vacuum cooker.

Steam jacketed pans are often fitted with scrapers and blades which make the mixing and heating process more uniform, and lessen the possibility of localized over-heating. Vacuum cookers are not generally used at a small scale.

Cooling

All sweets are cooled slightly before being shaped. Most simply, the boiled mass is poured onto a table (this should be made from metal, stone, or marble to cool the product uniformly). The table should be clean and free from cracks, as they may harbour dirt and micro-organisms.

It is important that the boiled mass is cooled sufficiently, since if it is to be formed by hand there is a danger that the operator may suffer burns.

	Mix ingredients	Boil	Cool	Beat	Form/set
Hard-boiled sweets	*	*	*		*
Fondant	*	*	*	*	*
Toffees/caramels	*	*	*		*
Fudge	*	*	*	*	*
Jellies	*	*	*		*
Marshmallows	*	*	*	*	*

Equipment required

Processing stage	Equipment	Section reference
Mix ingredients	Weighing and measuring equipment	64.1 64.2
Boil	Heat source Boiling pans Steam jacketed pans Thermometer	36.0 48.0 48.0 63.0
Cool	Table	
Beat	Hand whisk or liquid mixer	43.1
Form/set	Starch mould cutting equipment	17.1
Pack	Waxed papers cellulose films aluminium foils or polythene bags Heat sealer Wrapping equipment	47.1 47.3

Beating

Beating is a process which controls the process of crystallization and produces crystals of a small size. For example in the production of fudge, the mass is poured onto the table, left to cool, and then beaten with a wood or metal beater.

Forming/setting

There are two main ways of forming sweets: cutting into pieces, or setting in moulds.

Moulds may be as simple as a greased and lined tray. Other moulds can be made from rubber, plastic, metal, starch, or wood. It is possible to make starch moulds by preparing a tray of cornstarch (cornflour), not packed too tightly. Impressions are then made in the starch using wooden shapes. The mixture is poured into the impressions and allowed to set.

Boiling sugar syrup

Cooling marshmallows

Gutting gelatin sweets

Cutting toffee

Packaging

When sweets are stored without proper packaging, especially in areas of high humidity, the sucrose may crystallize, making the sweet sticky and grainy. Traditional packaging materials such as banana or sugar-cane leaves are often used to wrap sweets. However, these do not provide sufficient protection for a long shelf-life because they are not efficient barriers to moisture and cannot be securely sealed.

Alternatively, individual wraps can be made from waxed paper, aluminium foil, and cellulose film, or a combination of these. In most cases, the sweets will be wrapped by hand, but for higher production, semi-automatic wrapping machines are available. For further protection, the individually-wrapped sweets may be packed in a heat-sealed polythene bag.

Sweets can also be packaged in glass jars, or tins with close fitting lids.

Suitability for small-scale production

Certain types of sweets such as hard-boiled sweets require good quality ingredients (such as white granulated sugar). Such ingredients often need to be imported from other parts of the country, as they may not be widely or cheaply available in all areas.

It is possible to produce high-quality sweets on a small scale using inexpensive pieces of equipment. However, an open boiling pan gives less control over the boiling process and ultimately less control over the quality of the finished product. If simple equipment is used, the process will rely heavily upon the producer's experience and skills in production.

By using the basic principles of sweetmaking, the producer can use her/his creative skills to produce a wide range of products from local resources and materials.

7. Beverages

A wide range of plant materials are used to manufacture beverages. These include leaves, stems, sap, fruits, tubers, and seeds (grains). The large number of beverages may be classified as shown:

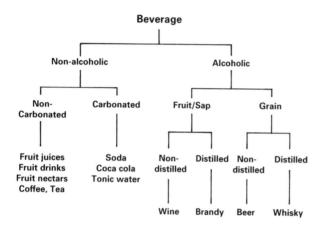

The market for beverages is broadly divided in many countries into those products that are bought to quench thirst, and those that are consumed on special occasions including festivals. The former group are mostly non-alcoholic and include tea, coffee, and soft drinks (including juices, nectars, and carbonated drinks). In some countries these products are also used on social occasions, whereas in other areas alcoholic beverages are preferred (although soft drinks are also usually available). In most countries, the market for alcoholic and non-alcoholic drinks is specific with regard to religious and cultural taboos.

Competition from medium/large-scale producers is most acute for small-scale producers in beverage manufacture. Many large-scale producers promote their products by implying status in their consumption and spend considerable amounts on advertising and packaging. They may also have established sophisticated distribution systems and specific agreements with wholesalers and retailers. Thus beverage manufacture is one of the most difficult for small-scale producers to establish and succeed in.

Nutritional significance

Most beverages contain a great deal of water. This does not add many nutrients to the diet, but it does play an important role in maintaining body balance by preventing dehydration.

Beverages are not usually consumed for their food value, but many, particularly the fruit drinks, contain quite a high percentage of sugar and therefore add to the energy content of the diet. Additionally fruit juices provide a supply of vitamins and minerals.

Certain drinks contain artificial flavourings and colourings. The use of such additives is governed by legal requirements and it is vital to keep to these regulations in order to protect the consumer from any undesirable side-effects. Some colouring agents for example are thought to cause hyperactivity in children, and are therefore to be avoided.

Alcoholic drinks are judged in terms of flavour and the stimulant effect they produce. In many countries alcohol production is strictly controlled by government agencies and it may be difficult to obtain the necessary permits to produce these beverages legally.

Non-alcoholic beverages

A wide range of drinks can be manufactured which contain as the base material either pulped fruit or juice. Many are drunk as a pure fruit juice without the addition of other ingredients, whereas others are diluted with sugar syrup.

For simplicity, fruit drinks can be divided into two groups:

○ Those that are drunk immediately after opening.

○ Those that are used little by little from bottles which are stored between use.

The former group should not need any preservative if processed and packaged properly. However the latter must contain a certain amount of permitted preservatives to have a long shelf-life after opening.

The following list may prove helpful in distinguishing between the different types of drink:

Juices. These are pure fruit juice with nothing added.
Nectars. These normally contain 30 per cent fruit solids and are drunk immediately after opening.
Squashes. These normally contain at least 25 per cent fruit pulp mixed with sugar syrup. They are diluted, to taste, with water and may contain preservatives.
Cordials. These are crystal-clear squashes.
Syrups. These are concentrated juices which are clear. They normally have a high sugar content.

Each of the above products is preserved by its natural acidity and by pasteurization. Some drinks (syrups and squashes) also contain a high concentration of sugar which helps to preserve them.

Alcoholic drinks

The most common examples of alcoholic beverages are wines and beers. Beer is usually made from a cereal, whereas wine can be produced from either cereals or fruit. Both can be distilled to produce spirits with an alcohol content of 30–50 per cent.

Both wines and beers are produced by fermentation which involves the conversion of sugars in the raw material or added sugar into alcohol and carbon dioxide. Different varieties of the yeast *Saccharomyces cerevisiae* are used to produce wines or beer. A simplified list of the differences is shown in the table opposite:

Product	Type of yeast
Beer	'Top yeast' – *Saccharomyces cerevisiae*
Lager	'Bottom yeast' – *Saccharomyces carlsbergensis*
Fruit wine	'Wine yeast' – *S. oriformis, S. chevalieri S. cerevisiae* (variety *ellipsoideus*) or a mixture of these
Palm wine	'Wild yeast' – natural mixture of yeasts
Rice wine	*Saccharomyces sake*

Although it is possible to use any strain of brewer's yeast for fermentation, it is necessary for a small producer to select one that works well and then continue to use it to produce a consistent product.

Alcohol has a lower boiling-point than water and distillation (vaporizing the alcohol and then condensing it) is used to concentrate the alcohol in spirit drinks. Distillation is carried out in stills which can be purchased for production at all levels. Alternatively, it is possible to construct a basic still using locally-available materials (see below).

A well-cleaned oil drum is fitted with a pipe to carry away the vapour, and a safety pipe. Alcoholic liquor is placed inside the drum and heated. On vaporization, the alcohol vapour is carried out of the drum via the pipe and passed through cooled air or cool water. The distillate condenses and is collected.

Distillation is more frequently carried out on a centralized commercial level, although it does occur on a small scale.

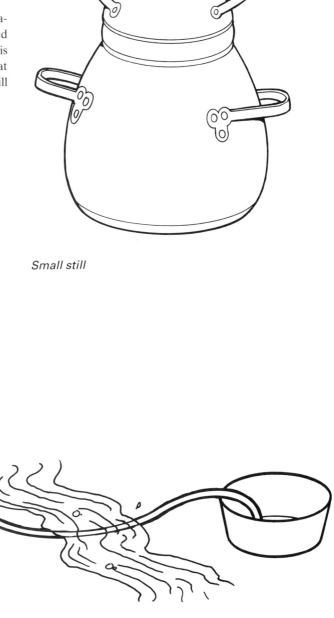

Small still

Traditional still

Beverages

	Juice extraction/ pulping	Mashing	Mix	Heat	Ferment	Filter	Bottle	Carbonate	Pasteurize
Beer		*	*	*	*	*	*	*optional	*
Wine	*		*	*	*	*	*		
Sparkling wine	*		*	*	*	*	*		
Fruit juice	*					*	*		*
Carbonated drink			*				*	*	*

The table above outlines the processing steps for the production of a range of representative beverages.

Equipment required

Processing stage	Equipment	Section reference
Juice pulping/ extraction	Fruit press Pulper/juicer	53.1 55.1 and 55.2
Mashing	Fermentation bins	03.1
Mixing	Mixers	
Boil	Boiling pans	48.1 48.2 48.3
Fermentation	Fermentation bins/jars	03.1
Filter	Filters and filter presses Sieves Strainers	29.1 and 29.2 29.3 29.4
Carbonation	Carbonating equipment	06.0
Filling into bottles	Liquid fillers Funnel	28.1
Pasteurize	Open boiling pan Steam jacketed pan Pasteurizer	48.1 and 48.2 48.3 50.0

Processing

Pulping/juice extraction

Either the juice or the pulp from fruit is the starting material for the manufacture of soft drinks and wines.

Pulping

Soft fruits, such as papaya, can easily be pulped by hand or by using a pestle and mortar. A wide range of hand-operated pulpers are available, or if electric power is available, multi-purpose kitchen-scale equipment such as blenders can be used. At an industrial level, this process is normally carried out in pulpers which brush the fruit through a sieve and eject the skin and stones. Smaller models of this machine can be manufactured and are commercially available.

Extraction

Juice can be extracted from fruit in several ways.

○ With a fruit press, fruit mill or hand pulper/sieve.

○ By crushing/pulping with a mortar and pestle and then sieving through muslin cloth or plastic sieves.

○ By steaming the fruit.

○ Citrus fruit juices need to be extracted by reaming (squeezing) the fruit, and once again, comparatively simple equipment is available for this purpose.

Fermentation

As mentioned previously the process for achieving fermentation differs considerably depending upon the product. The following paragraphs describe the basic processes for the production of both beer and wine.

Beer

The process for making beer is often referred to as brewing. Brewing actually consists of three stages – mashing, boiling and fermentation.

Mashing involves the use of hot water (approximately 68°C) to extract the soluble materials from the malted grains. This produces a liquid called wort. The process is carried out in large vessels which may be made of wood or stainless steel.

The wort is then subjected to a process of boiling. In Europe this process involves the addition of hops. Boiling takes place in a similar vessel to the tubs used for mashing except that it is flask-shaped, with the neck being elongated in order to carry away the steam, and to prevent over-boiling.

Prior to inoculation (addition of the yeast), the wort is cooled. This is because if added to the hot wort the yeast would be inactivated. The degree to which the wort is cooled differs according to the type of beer to be produced. For example, fermentation for lager is conducted at 12–15°C using the yeast *Saccharomyces carlsbergensis*. In other cases, the yeast *Saccharomyces cerevisiae* is used at a temperature of 20°C. During fermentation, the beer is held in fermentation vats or food-grade plastic fermentation bins. When fermentation is complete, the process of packaging will depend

Pulping juice

on whether the beer is to be sold in draught form (e.g. in a keg), or if it is to be bottled and corked. If it is to be draught, the beer is not filtered and small amounts of yeast are left in it in order to keep it slightly carbonated. In the case of bottled beer, it is filtered and pasteurized.

Wine

In wine-making, the fruit juice or pulp is mixed with yeast and sugar and held in a fermentation bin. Again this may be made from food-grade plastic. This is left for about ten days during the first fermentation stage. Within 48 hours, fermentation becomes vigorous and there is frothing and foaming. It is important to keep the fermentation vessel closed to prevent bacteria and fungi from infecting the wine. After ten days the fermenting wine is racked. This is done by scooping it up together with the solids using a sterilized mug, cup, or jug, and passing it through a muslin or nylon straining cloth. The cloth should have been sterilized and rinsed beforehand, and placed in a funnel. The wine is transferred into narrow necked fermentation vessels. These may be plugged with wads of cotton wool, or specially-designed vessels fitted with a airlock (known as a demijohn) may be used.

Ideally, fermentation is then continued at a temperature of 18°C. The whole process can take from three weeks to three months. The end of fermentation can be judged when

it is seen that there are no more bubbles rising to the surface. At this stage, the wine is filtered, in order to remove the sediment from the wine and then syphoned into narrow-necked or food-grade plastic vessels, and stored for the minimum period in the recipe to allow the wine time to clear and mature before bottling. After this period of maturation, the wine is siphoned off into bottles and sealed with a sterilized cork-stopper or screw cap.

Carbonation

This involves the addition of carbon dioxide into a drink. The most usual way of achieving this is to use a pressurized cylinder or tank which contains a mixture of water and carbon dioxide. In the case of soft drinks, the bottle is filled to a certain level with the flavoured syrup, the bottle is positioned under the cylinder head and carbon dioxide is released. The bottles are capped immediately. Cylinders for holding carbon dioxide are available for both large-scale production and in smaller sizes for use at the household level.

Pasteurization

Liquid products such as drinks may need to be pasteurized if they are to have a shelf-life of more than a few days. Pasteurization involves heating the product to a temperature of 80–90°C and holding it at that temperature for between

0.5 and 5 minutes before filling into clean sterilized bottles. Pasteurization is best carried out over a direct heat in stainless steel pans.

Some products can be pasteurized in their bottles. The filled bottles, with the lids loosely closed, are stood in a large pan of boiling water with the water-level around the shoulder of the bottle.

The time and temperature required for pasteurization will depend on the product and the bottle size.

Packaging

Beverages have differing needs with regard to storage, but the most pressing need for all beverages is simply that they need to be contained without the possibility of leakage.

The tables below outline some of the other storage requirements and the suitability of different types of container.

	Light	Air	Heat	Micro-organisms	Insects
Fruit juice, cordial etc.	some	*		*	*
Beer	*	*	*	*	*
Wine	*	*		*	*
Soft drinks				*	*

	Glass bottle/jar	Metal can	Plastic film/pot/ pouch	Ceramic pot
Fruit juice cordials etc.	*	Lacquered	*	*
Beer	Coloured	Lacquered	*	*
Wine	*			*
Soft drinks	*	*		

Glass bottles are the most popular medium for packaging beverages. However, owing to the expense of new glass, many producers (particularly those operating on a small scale) re-use the bottles. This means that in order to prevent contamination the bottles must be sterilized and cleaned properly. Simple hand-held bottle-brushes can be used to ensure a good standard of cleanliness, and mechanized brush-cleaners are also available.

Most beverages are thin liquids and can be filled quite easily by hand, but this is often too slow for a small business. A simple filler can be made by fitting one or more taps to the base of a bucket (see diagram below).

The bucket should be made from stainless steel for hot acid liquids (e.g. fruit juices) or food-grade plastic for cold filling. Iron and copper should not generally be used in food handling.

The type of closures used depends upon the type of product and its particular use (e.g. for glass bottles does it need to withstand internal pressure from carbonation).

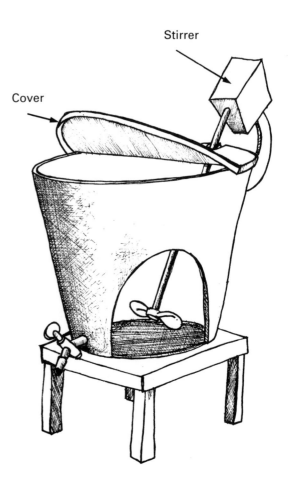

Liquid filler

There is a large range of closures available for glass bottles, but the choice for small-scale producers may often be restricted by what is locally available.

Metal 'crown' caps are commonly used for beers and fruit juices, whereas squashes, carbonated drinks and spirits are more frequently packaged using re-sealable metal screw-caps.

Wine is often sealed with a cork but plastic stoppers are equally effective and cost less.

With technological advances in the field of packaging materials, larger commercial manufacturers are using formed waxed cartons for beverages such as fruit juice. These have become very popular since they are cheaper and more convenient. Unfortunately the cost of the equipment needed to form and seal the cartons is very expensive and is presently out of reach for the small-scale producer.

Cheaper alternatives include plastic or foil laminated pouches. If sealed correctly, they can be a very convenient way of packaging.

Beverages can also be canned, but the cost of aluminium or steel cans is usually prohibitive for small-scale producers. In addition, the correct type of lacquer is required on the inside of the cans, and they are not re-usable.

Suitability for small-scale production

The manufacture of beverages is one of the most competitive areas in which small businesses can operate. Fruit drinks are the most accessible product (in technological terms) for small producers, but even with these there is strong competition from carbonated soft drinks, and it is necessary to establish that there is a demand for a certain drink before production starts.

Wine production is possible and can be very successful in some regions, provided there are not too many government restrictions on alcohol production. In most countries, beer and spirit production is dominated by large-scale, centralized producers, and it is very difficult for smaller producers to compete with them effectively. In some areas, however, there may be scope for upgrading traditional beers and spirits, by producing a uniform-quality product which is attractively packaged.

8. Vegetable oil

There is a universal demand for vegetable oil due to its use in domestic cooking, as an ingredient for other food production (in baked goods and fried snack foods), and as a raw material for the manufacture of soap, body/hair oils, and detergents.

Major oils and their uses

Raw material	Oil content (per cent)	Use
Oilseeds		
Castor	35–55	Paints, lubricants
Cotton	15–25	Cooking oil, soapmaking
Linseed	35–44	Paints, varnishes
Niger	38–50	Cooking oil, soapmaking, paint
Rape/mustard	40–45	Cooking oil
Sesame	35–50	Cooking oil
Sunflower	25–40	Cooking oil, soapmaking
Nuts		
Coconuts	64 dried copra 35 fresh nut	Cooking oil, body/hair cream, soapmaking
Groundnuts (peanuts)	38–50	Cooking oil, soapmaking
Palm kernel nuts	46–57	Cooking oil, body/hair cream, soapmaking
Shea nuts	34–44	Cooking oil, soapmaking
Mesocarp Oil palm	56	Cooking oil, Soapmaking

Oil can be extracted from many raw materials, but not all oil-bearing seeds, nuts, and fruit contain edible oil. Some contain poisons or unpleasant flavours and these are only used for paints; others such as castor oil need very careful processing in order to make them safe. Such oils are not suitable for small-scale processing.

Oils from certain crops such as maize are extracted by using solvents which dissolve the oil. This method of extraction is not suitable for small-scale operation due to the high capital costs of equipment, the need for solvents which may not be easily available, and the risk of fire or explosions.

Aflatoxins are a poisonous group of compounds produced by certain moulds which grow on seeds and nuts. Their occurrence has caused much concern recently, as they are poisonous to both humans and animals (causing liver damage, cancer, and death), if consumed over a prolonged period. The mould may grow either before or after harvest but this growth can be prevented (thereby preventing the production of aflatoxins) by drying the crop correctly. Aflatoxins are not destroyed or removed by heating, or during the subsequent processing stages and are difficult and expensive to remove if they occur – therefore grading is of particular importance in removing potentially dangerous produce. Seeds and nuts can be easily identified if there are visible signs of mould growth. There is, however, a danger that afflicted produce may not look mouldy after drying, and in these cases, produce contaminated with aflatoxins may only be detected by signs of discolouration or a shrivelled appearance.

If properly stored, vegetable oil has a shelf-life ranging from 6–12 months. Heat applied during processing destroys enzymes in raw materials, and also any contaminating micro-organisms which would cause rancidity. Additionally, the oil may be heated after extraction to remove as much water as possible. This lessens the occurrence of microbial spoilage during storage. Correct packaging and storage conditions slow down chemical changes caused by light and heat which may lead to rancidity.

Nutritional significance

Oil provides twice as much energy as the same quantity of carbohydrate and is therefore considered to be a valuable part of a well-balanced diet. Oil also contains a range of fat-soluble vitamins (A, D, E, and K) and essential fatty acids, both of which are necessary for the healthy functioning of the body.

The process of oil extraction produces a by-product known as oilcake. This is very nutritious, and can be used either for animal feed or as an ingredient in the production of other food products.

Processing

Oil is contained in plant cells, and its release depends on these cells being ruptured. Methods for achieving this depend on the composition of the raw material. For example, seeds, beans, and some nuts are processed in a dry state, whereas palm fruits are processed wet.

Principles of preservation:

○ to destroy enzymes in the raw material and contaminating micro-organisms by using heat during processing

○ removal of as much water from the oil as possible to prevent microbial growth during storage.

In addition, correct packaging and storage are important factors in preventing rancidity.

Girl collecting groundnuts

There are four main stages in the extraction of oil:

○ preparation of the raw material

○ extraction

○ clarification

○ packaging and storage

The following table outlines the stages involved in the processing of oils and also points out the type of equipment needed.

	Oilseeds (e.g. sunflower, safflower, mustard and sesame)	Groundnuts	Coconut (wet method)	Palm kernel	Palm fruit
Decorticate/ dehusk	*	*			
Crack			*	*	
Grind/grate		*	*	*	
Pulp					*
Heat/condition	*	*		*	
Press/expell	*	*	*	*	*
Clarify	*	*	*	*	*
Pack	*	*	*	*	*

Equipment required

Processing stage	Equipment	Section reference
Decorticate/ dehusk	Decorticators	19.0
Crack	Hand hammer Hammer mill Kernel cutters	41.3 17.3
Winnow	Winnowers	67.0
Pulp	Pestle and mortar or pulping machine	55.2
Grind/grate	Roller mill for groundnuts Hammer mill for palm kernels	41.2 41.3
Heat/ condition	Open pan for heating Heat source Thermometer Measuring and weighing equipment	48.0 36.0 63.0 64.1 and 64.2
Press/expell	Oil presses Expellers	53.2 26.0

Winnowing

Raw material preparation

Decortication
Some raw materials have a fibrous husk or seed coat, and this must be removed prior to processing. The removal is known as decortication and a range of decorticating machines are available which are suitable for small-scale production.

Winnowing
Winnowing takes place after decortication, and is the separation of the husks or seed coat from the oil-bearing material. Traditionally, this operation is achieved by gently throwing the seeds into the air and letting the air blow away the husks. This method of winnowing requires skill and experience. For higher rates of production, it is possible to use either a manual or a powered winnowing machine.

Cracking
Palm kernels and coconuts need to be cracked and the shell removed before processing can begin. This can be achieved manually using simple tools such as a hammer or a heavy knife. Motorized hammer mills are available for cracking but for coconuts it seems that the manual method is presently more efficient.

Pulping
Palm fruits can be pulped manually with a pestle and mortar, or more quickly, with a motorized pulper.

Grinding/flaking

Traditional hand-pounding methods using a pestle and mortar or more sophisticated roller mills, may be employed to grind groundnuts into a coarse flour. Flakers are used for sunflower seeds, and hammer mills are applicable for palm kernels. Coconut flesh needs to be grated, and a wide range of manual graters are available.

Heating/conditioning

Many raw materials, such as groundnuts and sunflower seeds, are heated with water prior to extraction. This processing stage is known as 'conditioning' and has the following effects:

○ it assists in the rupturing of the oil-bearing cells

○ it decreases the thickness of the oil, allowing it to flow more easily.

The desired moisture content will differ, according to the raw material being used. For example, groundnut flour needs an addition of 10 per cent water and is heated at a temperature of 90°C. The heating is traditionally carried out over open fires, although seed scorchers, which are basically pans fitted with stirrers, are now available to mix the nuts/seeds better.

Assessing the correct amount of moisture and degree of heating is often quite difficult, and therefore a simple hand-

Conditioning groundnut flour

feel test is used. For example, in the case of groundnuts, the heating process is complete when the mixture stops sticking together and forms a free-flowing flour again.

Oil extraction

One traditional method for extracting oil involves the use of a 'ghani'. Ghanis originated in India, but their use is now more widespread.

A ghani in operation

A ghani consists of a wooden mortar and pestle. The mortar is fixed to the ground, and the pestle is located in the mortar, where the raw material is crushed by friction and pressure. An animal is required to move the pestle and as this continues the oil is pressed out, runs through a hole at the bottom of the mortar, and the residue (cake) is then scooped out. Ghanis are limited in that two animals are required, since any one animal will tire after 3–4 hours.

Improved technologies for extracting oil include:

○ motorized ghanis

○ oil presses

○ oil expellers.

Motorized ghanis

These are becoming increasingly popular and are fast replacing animal-powered equipment.

Oil presses

Many types of oil press are available, but all work on a similar principle. Raw materials are placed in a heavy perforated or slotted metal cage and a metal plunger is used to press out the oil. The main differences in design are as follows:

○ The method used to move the plunger.

○ The amount of pressure in the press.

○ The size of the cage.

The plunger can be moved either manually or with the aid of a motor. The motorized method is faster but more expensive. Different designs use either a screw-thread or a hydraulic system.

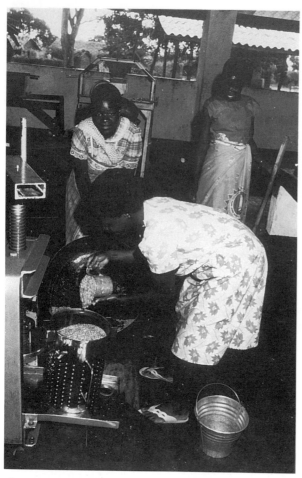

A manual oil press

In most cases, oil presses can be manufactured locally. If it is a screw-press, however, a lathe is needed to manufacture the screw. Hydraulic presses may be manufactured locally if lorry jacks are available. It is important that the mineral oil used with either the screw or the hydraulic press does not contaminate the oil.

Oil expellers

Expellers use a horizontally-rotating screw which feeds oil-bearing raw material into a barrel-shaped outer casing with perforated walls. The raw material is continuously fed to the expeller, which grinds, crushes, and presses out the oil as it passes through the machine. Oil flows through the perforations in the casing and is collected underneath. The residue, or oilcake, is pushed out of the end of the unit.

Most small expellers are power-driven. Due to the wear and tear, it is likely that the screw will need to be repaired and replaced at frequent intervals. Therefore, it is essential that the skills and resources required for maintenance are available locally.

Clarification

Crude oil contains a suspension of fine pulp and fibre from plant material. It also contains smaller quantities of water, resins, colours, and bacteria which makes it darker in colour. These contaminants are removed by clarifying the oil, either by allowing the oil to stand undisturbed for a few days and then removing the upper layer, or by using a clarifier. If further clarification is needed, the oil may be filtered

Tinytech oil expeller

through a plastic funnel which has been fitted with a fine filter cloth. Finally, the oil is heated to boil off the traces of water and destroy any bacteria. For those raw materials which are processed wet (such as coconut), heating is applied prior to clarification in order to break the emulsion. When these impurities are removed, the shelf-life of the oil can be extended from a few days to several months, provided it is stored properly.

Packaging and storage

Rancidity can cause the oil to deteriorate and develop 'off' flavours during storage. This may be prevented by using clean, dry containers which exclude light and heat, and prevent contact with metals such as iron or copper.

Sealed glass or plastic bottles are adequate packaging materials. It is preferable if bottles are made from coloured glass and kept in a dark box. Metal cans may be used provided the metal is tin-coated. Alternatively, glazed ceramic pots sealed with a cork and a wax stopper are also suitable.

Re-used bottles must be well-cleaned to ensure that there is no film of old rancid oil on the inside of the container. If not, it will quickly turn new oil rancid. The containers should be properly dried after cleaning to remove all traces of water.

If the oil is packaged adequately and kept away from heat and sunlight, the shelf-life can be expected to be 6–12 months.

Suitability for small-scale production

Only seeds, nuts, and fruits which contain considerable amounts of edible oil are used for small-scale extraction. Other types may contain edible oil but small-scale equipment cannot extract enough of the oil to be economically viable.

Oilseeds have a long shelf-life if kept sufficiently dry (thereby preventing the growth of mould and the subsequent occurence of aflatoxins), and oil processing can therefore continue throughout the year, and make better use of the equipment than more seasonal products.

The low volume of oil, compared to oilseeds, makes transport and distribution easier and more effective. Distribution to a wider area creates more potential markets, and producers are able to receive an income throughout the year.

The residual oilcake is often highly nutritious and can be used for either animal or human food. For example, the cake produced from peanut oil extraction can be fried and made into a snack food.

There are also disadvantages to oil extraction. These include the higher value of oilseeds compared to other crops, the higher financial risk from losses, the need for fuel, and the time-consuming hard work entailed. The equipment can be expensive, and year-round production needs a large working capital to buy and store the seasonal crops. There is also the risk of competition from large-scale producers who are able to market high-quality oils at a lower cost because of economies of scale.

Due to competition from larger-scale manufacturers and the importation of cheaper oils in some areas, it is often more profitable for small producers to use the oil as part of the secondary processing of other products, such as soap.

Small-scale extraction processes produce crude oil. This has a different appearance and flavour compared to commercially-refined oils. It is therefore necessary to test market the crude oil for acceptability before embarking on a processing enterprise.

9. Milk and milk products

Although cow's milk is the most popular in many countries, milk can be obtained from many different sources. For example, milk from goats and sheep makes a substantial contribution to the total milk production in countries of Eastern and Southern Europe, Malawi, and Barbados, whereas the water buffalo is a common source of milk in much of Asia. The table below illustrates some of the differences in composition between these milks.

Woman milking a goat

Milk is a perishable commodity and spoils very easily. Its low acidity and high nutrient content make it the perfect breeding ground for bacteria, including those which cause food poisoning (pathogens).

Bacteria from the animal, utensils, hands, and insects may contaminate the milk, and their destruction is the main reason for processing. This preservation of the milk can be achieved by fermentation, heating, cooling, removal of water, and by concentration or separation of components, to produce foods such as butter or cheese.

The degree to which milk consumption and processing occurs will differ from region to region. It is dependent upon a whole host of factors, including geographic and climatic conditions, availability and cost of milk, food taboos, and religious restrictions. Where processing does exist, many traditional techniques can be found for producing indigenous milk products. These are more stable than raw milk and provide a means of preservation as well as adding variety to the diet. In addition, the introduction of western-style dairy products and the subsequent setting up of small-scale dairies has provided more choice of dairy products to the consumer.

Nutritional significance

Milk is often regarded as being nature's most complete food. It earns this reputation by providing many of the nutrients which are essential for the growth of the human body. Being an excellent source of protein and having an abundance of vitamins and minerals, particularly calcium, milk can make a positive contribution to the health of a nation. The realization of its nutritional attributes is clearly illustrated by the implementation of numerous 'school milk programmes' worldwide.

Fermented-milk products such as yoghurt and soured milk contain bacteria from the *Lactobacilli* group. These bacteria occur naturally in the digestive tract and have a cleansing and healing effect. Therefore the introduction of fermented products into the diet can help prevent certain yeasts and bacteria which may cause illness.

Many people suffer from a condition known as 'lactose intolerance'. This means that they are unable to digest the milk fat (lactose). Such people can, however, tolerate milk if it is fermented to produce foods such as yoghurt. During fermentation, lactic acid producing bacteria break down lactose, and in doing so eliminate the cause of irritation.

The quality of milk

The type of animal, its quality, and its diet can lead to differences in the colour, flavour, and composition of milk. Infections in the animal which cause illness may be passed directly to the consumer through milk. It is therefore

Average composition (%) of milks of various mammals

Species	Water	Fat	Protein	Lactose	Ash
Human	87.43	3.75	1.63	6.98	0.21
Cow	87.2	3.7	3.5	4.9	0.7
Goat	87.00	4.25	3.52	4.27	0.86
Sheep	80.71	7.9	5.23	4.81	0.9
Indian buffalo	82.76	7.38	3.6	5.48	0.78
Camel	87.61	5.38	2.98	3.26	0.7
Horse	89.04	1.59	2.69	6.14	0.51
Llama	86.55	3.15	3.9	5.6	0.8

Schoolchildren receiving milk   *FAO photo*

extremely important that quality-control tests are carried out to ensure that the bacterial activity in raw milk is of an acceptable level, and that no harmful bacteria remain in the processed products.

Standard testing procedures

Milk fat

The price paid for milk is usually dependent upon the milk-fat content, and this may be determined either at the collection stage or at the dairy using a piece of equipment known as a butyrometer. Additionally the specific gravity can be measured using a hydrometer. This can also be used as an aid to detect adulteration.

Bacterial activity

Routinely it is necessary to check the microbiological quality of raw milk using either methylene blue or resazurin dyes. These tests indicate the activity of bacteria in the milk sample and the results determine whether the milk is accepted or rejected.

Both tests work on the principle of the time taken to change the colour of the dye. The length of time taken is proportional to the number of micro-organisms present (the shorter the time taken, the higher the bacterial activity). It is preferable to use the resazurin test as this is less time-consuming. For these tests, basic laboratory equipment will be needed such as test-tubes, a water bath, accurate measuring equipment, and a supply of dyes.

After collection the milk should ideally be stored at a temperature of 4°C or below. This is necessary to slow the growth of any contaminating bacteria.

Phosphatase test

For pasteurized milk, it is possible to ensure that pasteurization has been adequately achieved by testing for the presence of the enzyme phosphatase. The destruction of phosphatase is regarded as a reliable test to show that the milk has been sufficiently heat-processed, because this enzyme (present in raw milk) is destroyed by pasteurization conditions.

It is stressed that pasteurization is an effective safeguard against spoilage and food poisoning only if the milk is not re-contaminated after pasteurization.

Processing

Liquid milk

Milk can be kept for longer periods of time if it is heated to destroy the bacteria or cooled to slow their growth. Pasteurization and sterilization are the two most commonly-used heat treatments. Technically, it is possible for both to be carried out on a small scale, but they are most usually performed on a larger industrial scale due to the need for qualified, experienced staff and accurate and strictly controlled hygienic processing conditions.

Production stages for pasteurized and sterilized milk

Product	Store raw milk at 4°C	Test for bacterial activity using resazurin/ methylene blue	Filter	Homogenize	Pasteurize	Fill into sterilized bottles	Sterilize	Store
Pasteurized milk	*	*	*		*	*		*
Sterilized milk	*	*	*	*		*	*	*

Milk and milk products

Equipment required

Processing stage	Equipment	Section reference
Store at 4°C	Refrigerated storage	15.0
	Thermometer	63.0
Test for fat content	Butyrometer	64.5
Test specific gravity	Hydrometer	64.4
Test bacterial activity	Supply of dyes	64.6
	Thermometer	63.0
	Basic laboratory equipment is required for most of the tests	
Filter	Filter cloth	0.80
	Filter press	29.2
Homogenization	Homogenizer	37.0
Fill into bottles	Liquid-filling machine	28.1 – refer to Packaging chapter for notes on the preparation of sterilized bottles
	Capping machine	47.2
Pasteurization	Boiling pan	48.0
	or pasteurizer	50.0
	Heat source	36.0
	Thermometer	63.0
Sterilization	Pressure cooker	48.0
	Retort	05.1
	Heat source	36.0
	Thermometer	63.0
Cool	Bottle-cooling system	Refer to the Packaging chapter for details

Homogenization

Homogenization breaks up the oil droplets in milk and prevents the cream from separating out and forming a layer. This is of particular importance for sterilized milk which has a long shelf-life and when the formation of a cream layer is not desired. Additional changes include increased viscosity and a richer taste. Homogenizers are more usually designed for industrial-scale production, but it is possible to purchase smaller versions.

Filling

The most common packaging material for both pasteurized and sterilized milk is glass bottles sealed with either foil or metal caps, although plastic bottles, plastic bags, and cardboard cartons are all used when bottles are not available or too expensive.

Pasteurization

Pasteurization is a relatively mild heat treatment,(usually performed below 100°C) which is used to extend the shelf-life of milk for several days. It preserves the milk by the inactivation of enzymes and destruction of heat-sensitive micro-organisms, but causes minimal changes to the nutritive value or sensory characteristics of a food. Some heat-resistant bacteria survive to spoil the milk after a few days, but these bacteria do not cause food poisoning.

The time and temperature combination needed to destroy 'target' micro-organisms will vary according to a number of complex inter-related factors. For milk, the heating time and temperature is either 63°C for 30 minutes or alternatively 72°C for 15 seconds. Only the former combination is possible on a small scale and for this the simplest equipment required is an open boiling pan. Better control is achieved using a steam jacketed pan, and this can be fitted with a stirrer to improve the efficiency of heating. Both of these are batch processes which are suited to small-scale operation. A higher production rate may be possible using a tubular-coil pasteurizer. This equipment has been tested and has been successful for some fruit products but it is presently still at a developmental stage.

Sterilization

Sterilization is a more severe heat treatment designed to destroy all contaminating bacteria. The milk is sterilized at a temperature of 121°C maintained for 15–20 minutes. This can be achieved using a retort or pressure cooker. Unlike pasteurization, this process causes substantial changes to the nutritional and sensory quality of the milk. In some countries, flavoured milk has become a very popular product.

However, sterilization is not recommended for small-scale production for the following reasons:

○ The cost of a retort and ancillary equipment is high for the small-scale processor.

○ It is essential that the correct heating conditions are carefully established and maintained for every batch of milk that is processed. If the milk is overheated, the quality is reduced, and it may have a rather burnt taste and aroma.

○ If the milk is not heated sufficiently, there is a risk that micro-organisms will survive and grow inside the bottle. In low-acid foods such as milk, many types of bacteria including *Clostridium botulinum* can grow and cause severe food poisoning.

○ Due to the potential dangers from food poisoning, the skills of a qualified food technologist/microbiologist are required in order to routinely examine samples of sterilized milk that have been subjected to accelerated storage conditions. This requires a supply of microbiological media and equipment.

In summary, the process of sterilization requires a considerable capital investment, the need for trained and experienced staff, regular maintenance of sophisticated equipment, and a comparatively high operating expenditure.

Cooling

Pasteurization does not destroy all of the micro-organisms, therefore the milk has to be cooled rapidly to prevent the growth of surviving bacteria. Cooling can be achieved on a small scale by using a bottle-cooling system. A system is outlined in the Packaging chapter.

Storage

Pasteurized milk has a shelf-life of 2–3 days if kept at 4°C. Maintaining this low temperature causes a substantial increase to the cost of transportation and distribution and is therefore a major disadvantage to the development of a small-scale pasteurized milk business. If packaged in sealed bottles and stored at room temperature, sterilized milk should have a shelf-life in excess of six months.

Separation of milk components

Cream

When milk is left to stand for some time, fat globules rise to the surface forming a layer of fat (or cream). This can be separated leaving behind skimmed milk as a by-product. There are different types of cream each with different fat concentrations: single (or light) cream contains 18 per cent milk fat whereas double (or heavy) cream normally contains 30 per cent milk fat. Cream is a luxury item and may be used as an accompaniment to coffee, as a filling in cakes, and an ingredient in ice cream.

Separation

Separation can very simply be achieved by removing the cream with a spoon, however this is a slow process during

which the cream may spoil. For this reason it is more usual to use a manual or powered centrifuge.

Pasteurization

Cream may be pasteurized in a similar way to milk, using a similar time and temperature combination and the same equipment. Cream can also be sterilized but there is a considerable loss of quality.

Packaging and storage

Cream can be packaged in glass jars or plastic pots sealed with foil lids. Pasteurized cream must be stored at a temperature of 4°C to have a shelf-life of several days. Refrigerated storage is necessary because cream is prone to rapid spoilage.

Butter

Butter is a semi-solid mass which contains approximately 80–85 per cent milk-fat and 15–20 per cent water. It is yellow/white in colour, with a bland flavour and a slightly salty taste. It is a valuable product that has a high demand for domestic use in some countries and as an ingredient in other food processing (e.g. for confectionery and bakery uses).

The principles of preservation are :

○ to destroy enzymes and micro-organisms by pasteurizing the milk

○ to prevent microbial growth during storage by reducing the water content, by storing the product at a low temperature, and optionally by adding a small amount of salt during processing.

Production stages for cream

Ingredients	Process	Equipment	Section reference
Raw milk tested	Store at 4°C	Milk churns Refrigerated storage Thermometer	62.0 15.0 63.0
	Separation of milk fat	Ladle Dairy centrifuge	07.1
	Pasteurization	Large boiling pan or steam jacketed pan Pasteurizer Heat source Thermometer	48.0 50.0 36.0 63.0
	Fill bottles/ pots	Funnel or liquid-filling machine Capping machine Pot sealer	28.1 47.2 47.1
	Cool bottles	Bottle-cooler	See Packaging chapter
	Store bottles at 4°C	Refrigerated storage	15.0

Production stages for butter

Ingredients	Process	Equipment	Section reference
Cream or soured cream	Store at 4°C	Milk churns Refrigerated storage Thermometer	62.0 15.0 63.0
	Churning	Butter churns	13.0
	Draining (pour off buttermilk)		
	Washing		
	Draining (pour off washwater)		
Permitted colours and salt (optional)	Kneading/ working	Butter pats	04.0
	Form into blocks	Butter pats	04.0
	Packaging	Paper/plastic/ foil wrapping Wrapping machines	47.3
	Storage at 4°C	Refrigerated storage	15.0

Churning

Churning disrupts the emulsion of fat and water and as a result the milk-fat separates out into granules. This process takes place in a butter churn.

Churning cream

Churning is continued until fat granules are present and at this stage the mixture is drained to remove liquid that has separated from the granules. This liquid is known as buttermilk and can be used as either a beverage or as an ingredient in animal feed.

Washing

Clean water equivalent in weight to the buttermilk is added to the churn in order to wash the butter granules. The wash water is drained away. Churning is continued for a short time to compact the butter, and once this has been achieved it is removed from the churn.

Forming and packaging

Butter is kneaded to achieve a smooth and pliable texture. This can be done using simple hand-tools such as butter pats. Alternatively for higher production rates a specially-designed kneader can be used. Once the butter has a uniform and smooth texture it is formed into blocks with butter pats and packed in either greaseproof paper or foil wrappers.

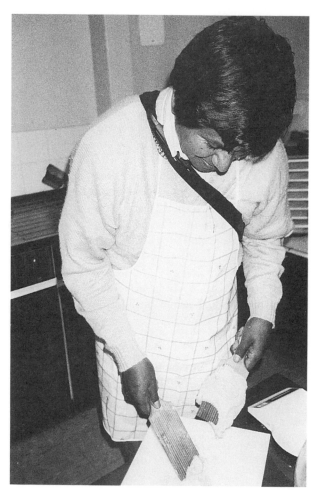

Working butter with butter pats

Storage

Due to its high fat composition, butter must be stored at temperatures below 10°C otherwise the fat becomes rancid and imparts undesirable 'off' flavours. The water droplets in butter (20 per cent) can also allow bacteria to grow if it is not kept under cool conditions.

Ghee

Ghee is made from butter which has been heated and clarified. At ambient temperatures it is a semi-solid mass with a granular texture, but on melting (40°C+) it turns into a clear, thin liquid. It has a high demand in some countries for domestic use, as an ingredient for local food production (for example bakeries and confectionery manufacturers), and as an export commodity.

Alternatively, cream is boiled gently to evaporate the water. During boiling the product is stirred continuously until the milk proteins start to coagulate, forming particles, and the colour of the cream darkens. Heating is stopped and the product is left to set. The particles settle at the bottom of the vessel and the milk-fat is separated.

The principles of preservation are:

○ heating to destroy enzymes and contaminating micro-organisms

○ to reduce the water-content by evaporation, and in doing so prevent the growth of micro-organisms.

Production stages for ghee

Ingredients	Process	Equipment	Section reference
Butter	Heating	Heat source Large boiling pan or steam jacketed pan	36.0 48.0
	Cool to room temperature	Thermometer	63.0
	Filter	Filter cloth	08.0
	Fill into jars/ pots	Funnel or liquid-filling machine Capping machine	28.1 47.2
	Store at ambient temperatures		

Packaging and storage

Metal containers are normally used. They should be thoroughly cleaned, especially if they are re-usable, and they should be made airtight. Alternatives to metal cans include coloured glass jars with metal lids, or ceramic pots sealed with cork/plastic stoppers.

Ghee is usually stored at room temperatures as cold storage affects the granular texture. Thus ghee is useful for those consumers with no access to refrigeration.

Cultured/fermented dairy products

The technology of cultured milk products such as yoghurt, curd, and cheese is based upon the microbial conversion of the milk-sugar lactose to lactic acid (lactic acid accounts for the characteristic 'sourness' of such products). In order for the conversion to take place, lactic acid producing bacteria must be present. This may occur by allowing the milk to sour naturally, but it is better to introduce the appropriate bacteria as a starter culture. Starter cultures may be in the form of a small quantity of previously-cultured product or may be purchased as a commercially-prepared culture.

Yoghurt/curd

Yoghurt is a fermented milk product that evolved by allowing naturally-contaminated milk to sour at a warm temperature. Yoghurt can be either unsweetened or sweetened, set, or stirred. Curd is the name given to a yoghurt-type product made from buffalo milk.

The principles of preservation for yoghurt are:

○ Pasteurization of the raw milk to destroy contaminating micro-organisms and enzymes.

○ An increase in acidity due to the production of lactic acid from lactose. This inhibits the growth of food-poisoning bacteria.

○ Storage at a low temperature to inhibit the growth of micro-organisms.

Production stages for set yoghurt

Ingredients	Process	Equipment	Section reference
Milk and starter culture (2 per cent)	Preheat to 70°C for 15–20 minutes	Heat source Thermometer Boiling pan	36.0 63.0 48.0
	Cool to 30–40°C	Thermometer	63.0
	Addition of starter culture	Measuring and weighing equipment	64.1 and 64.2
	Pour into bottles/pots	Funnel or Liquid filler Sealing machine or Capping machine	28.1 47.1 47.2
	Incubate at 43–45°C	Commercial incubator Thermometer	39.0 63.0
	Store at 4°C	Refrigerated storage	15.0

Heating

In the manufacture of yoghurt, milk is normally heated to 70°C for 15–20 minutes, using an open boiling pan, or alternatively a steam jacketed pan.

Addition of starter culture

The milk is cooled to between 30–40°C and inoculated with a mixed culture of *Lactobacillus bulgaricus* and *Streptococcus thermophilus* (usually in a ratio of 1:1). If a commercial starter-culture is used, the directions for use will be given. However, if a culture from a previous batch is used, then it is usual to add 2-3 tablespoons per litre of prepared milk.

Yoghurt of the stirred variety can be fermented in the mixing container. To make set yoghurt the inoculated milk should be poured into the individual pots before fermentation.

Selling curd from a roadside stall

Incubation

The micro-organisms that produce yoghurt are most active within a temperature range of 32–47°C. Ambient temperatures are therefore not adequate and a heated incubator is needed. Small commercially-available yoghurt-makers consist of an electrically-heated base and a set of plastic or glass containers. Most yoghurt-makers make four or five individual half litre cups at a time. There are other simple and inexpensive ways of incubating yoghurt such as an insulated box, keeping the jars/pots surrounded by warm water, or by using thermos flasks (the latter is only suitable for stirred yoghurt). Incubation takes approximately five hours.

When fermentation is complete, stirred yoghurt is cooled and flavoured or sweetened prior to packaging. In set yoghurt all ingredients are added before fermentation.

Packaging and storage

Yoghurt or curd is commonly packaged in plastic pots fitted with a plastic lid, or heat-sealed with foil, although traditionally, curd is packaged in clay pots. Such pots are made from local materials and can be re-used or later used for cooking. The shelf-life of yoghurt is usually 3–8 days when stored at temperatures below 10°C.

Cheese

Cheese is made from milk by the combined action of lactic acid bacteria and the enzyme rennin (known as rennet). Just as cream is a concentrated form of milk fat, cheese is a concentrated form of milk-protein. The differences in

Collecting milk for cheese-making

cheeses that are produced in different regions result from variations in the composition and type of milk, variations in the process, and the bacteria used. The different cheese varieties can be classified as either hard or soft.

Hard cheeses such as Cheddar and Edam have most of the whey drained out and are pressed. Soft cheeses such as paneer contain some of the whey and are not pressed. Many indigenous cheeses are soft types.

The hardness, flavour, and other qualities of a cheese can be varied by changes to the process conditions, to suit local tastes. However the principal steps of a cheese-making process are basically the same.

The principles of preservation are:

○ the raw milk is pasteurized to destroy most enzymes and contaminating bacteria

○ fermentation by lactic-acid bacteria increases the acidity which inhibits the growth of food-poisoning and spoilage bacteria

○ the moisture content is lowered and salt is added to inhibit bacterial and mould growth.

The table, right, outlines the stages of production and the equipment needed to produce Edam cheese.

Pre-heating
The pasteurized milk is heated to a temperature at which the starter-culture can work.

Addition of starter culture
Starter-culture is added to the milk at the rate of approximately 2 per cent of the weight of milk. The vessel used should be either aluminium or stainless steel.

Addition of rennet
The rennet should be 1 per cent of the weight of milk. The rennet alters the milk proteins and allows them to form the characteristic curd.

Incubation
The milk is allowed to stand until it sets to a firm curd.

Treatment of the curd
The curd is cut into cubes which facilitate the elimination of whey from the gel. The curd is then cooked at 40°C for a period of twenty minutes which has the action of firming the curd. After cooling, the whey is drained off. The curd is pressed to ensure that most of the whey has been removed, and is then cut to fit the cheese-moulds, and finally pressed with weights.

Ripening
This is the final stage in the cheese-making process. It is a process which allows the development of gas in some cheeses and the development of flavour. The longer the ripening process the stronger the flavour. Ripening usually takes place in ripening rooms, where the temperature and humidity must be controlled for the optimum development of the cheese.

Packaging and storage
The packaging requirements differ according to the type of cheese produced. Hard cheese, for example, has an outer protective rind which protects the cheese from air, microorganisms, light, moisture-loss or pick-up, and odour pick-up. Cheese should be allowed to 'breathe', otherwise it will sweat. Suitable wrapping materials are therefore cheese-cloth or grease-proof paper. Cheese should be stored at a relatively low temperature between 4 and 10°C to achieve a shelf-life of several weeks/months. Soft cheeses are often stored in pots or other containers, often in brine, to help increase their shelf-life of several days/weeks.

Production stages for Edam-type cheese

Ingredients	Process	Equipment	Section reference
Pasteurized milk	Preheat to 35–40°C	Cheese vat or boiling pan Thermometer Heat source	10.0 48.0 63.0 36.0
Starter culture	Addition of starter culture	Measuring and weighing equipment	64.1 and 64.2
Rennet	Add rennet at 30°C	Measuring and weighing equipment Thermometer	64.1 and 64.2 63.0
	Incubate		
	Cut the curd	Curd cutters	16.1
	Heat to 40°C for 20 minutes	Heat source Thermometer	36.0 63.0
	Drain	Filter cloth	08.0
	Cut to fit a cheese-mould	Knife	17.1
	Put into a cheese-mould	Cheese-moulds	09.1
	Press with weights	Cheese-press	09.2
	Cool and dry at 10–12°C	Thermometer	63.0
Salt	Salting in 20 per cent salt solution at 12°C for 12–16 hours	Brine meter Thermometer	64.6 63.0
	Ripen for 6–8 weeks at 16°C	Thermometer (optional)	63.0
	Washing Drying for 30 minutes Wax with paraffin wax store at 9°C	Refrigerated storage	15.0

Selling ice cream from a bicycle

Ice cream

Ice cream is a frozen mixture which contains milk, sugar, fat, and optional thickeners (e.g. pectin or gelatin), colouring, and flavouring. It may be sweetened and flavoured in numerous ways with nuts, fruit pieces, and natural or artificial flavours and colours.

The principles of preservation are:

○ pasteurization to destroy most micro-organisms and enzymes

○ freezing to inhibit microbial growth.

Pasteurization

Pasteurization is carried out by heating to 65°C for a period of 30 minutes.

Cooling and beating

Ice cream is a complex mix of small ice crystals and air bubbles in a milk-fat/water emulsion. To achieve this, it is necessary to cool the mixture quickly to produce small ice crystals and at the same time incorporate air into the product by beating.

Ingredients	Processing stage	Equipment	Section reference
Milk, sugar, fat, thickener, colours and flavours.	Mix ingredients	Weighing and measuring equipment	64.1 and 64.2
		Liquid mixer	43.1
	Pasteurize	Boiling pan or steam jacketed pan	48.0
		Thermometer	63.0
		Heat source	36.0
	Cool mixture to approximately −5°C and beat simultaneously	Ice cream maker	38.0
	Fill into containers	Filling machine	28.0
	Freeze at −18°C	Freezer	32.0

Ice cream makers are available commercially and work on the following principle. The mixture is placed in a bowl which is kept at a low temperature (either surrounded by ice and salt, or having been chilled in a freezer). It is then agitated by a manually-operated rotor or by a powered stirrer. At the end of this process the ice cream should be at a temperature of approximately –5°C, and be partly frozen.

Packaging and storage

The ice cream is usually packaged in plastic, waxed paper, or cardboard containers, and is stored at below –18°C. The storage temperature is important for two reasons:

○ to maintain the texture of the product

○ to prevent the growth of micro-organisms.

Ice cream may be transported in an insulated box (e.g. for sale from a bicycle). It is especially important to guard against thawing and re-freezing as this will cause changes in texture and mouthfeel, and there is the increased possibility of food poisoning by contaminating food poisoning micro-organisms.

Suitability for small-scale production

In some regions, there is a high demand for dairy products – both for traditional and modern items. Much of the technology and machinery necessary for processing is fairly simple, which may at first sight appeal to the would-be processor. Milk, however, is a highly perishable food and there is a high risk of transmitting food-poisoning bacteria to consumers. It is stressed, therefore, that milk processing of any kind must be done under carefully controlled hygienic conditions.

Furthermore, in developing countries, milk processing is often more problematic than in temperate climates, owing to higher temperatures and humidity. Consequently, milk spoils at a faster rate, cheese ripens too quickly, and it is often difficult to ensure adequate cooling conditions.

10. Meat and meat products

Any animal can be used as a source of meat, ranging from the domesticated cow, pig, and chicken, to deer and camels.

The amount and type of meat eaten in a particular country is determined by factors such as cost, availability, and cultural or religious acceptability. In developing countries, meat is usually eaten fresh, and meat products, with the exception of dried meat in some African countries, are not commonly available or in significant demand.

Conditions under which animals are slaughtered in developing countries are often unhygienic, and animals may be slaughtered and displayed for sale in the open air without any covering. This enables bacteria to be transmitted by flies, other animals, and birds. Meat, like fish and milk, is a low-acid, moist food which provides a good environment for the growth of these bacteria. This may lead to rapid deterioration of the meat in tropical climates and fresh meat therefore has a very short shelf-life. Harmful micro-organisms may also grow on meat and produce food poisoning when eaten. This, together with infectious organisms such as parasites which grow in the meat, make proper handling and preparation of meat products essential.

In this chapter the focus is on processed meat products and not the sale of fresh meat for direct consumption.

However, fresh meat is the raw material for processing, and a note on hygiene and handling is therefore included below.

Unlike other commodity groups in this book, there are a limited number of products made from meat. For this reason, only a few examples which may be suitable for production on a small scale, such as sausage, dried meat, and burger patties, are included. It is strongly suggested that meat products should only be considered by experienced food processors, because of the potential risks from food poisoning.

The two main methods of processing meat products are first, salting, smoking and/or drying, and second grinding to form minced meat, sausage-meat or pâté.

In this chapter, one example from each group is described ('biltong' as an example of a dried, salted meat, and sausages as an example of a ground (or comminuted) meat product). It should be noted also that there are hundreds of types of sausages, many of which are also smoked and/or dried to aid preservation. Freezing and canning are techniques not covered in this chapter. Both are excluded as they are capital-intensive operations and both have a serious risk of food poisoning which is thought to be unacceptable for small-scale producers.

A meat stall

Nutritional significance

Meat is a good source of easily digestible protein and contains essential amino acids which are vital for growth and maintenance of the body. It is also a good source of vitamins and minerals particularly iron. Processing, to form sausages, patties, or dried meat, does not have a substantial effect on the nutritive value when compared to normal cooking processes.

Meat processing

Fresh meat should be kept under refrigeration or cool storage and covered to protect it from insects and animals. Additionally, the hands and clothes of workers who handle meat should be regularly cleaned.

Salted, smoked, and dried meat

In all meat processing, the aim is twofold: to preserve the meat for a longer storage life, and to change the flavour and texture to increase variety in the diet. In smoking, the effects of heat from the smoke, and chemicals in the smoke, combine to preserve the meat. Smoke also adds distinctive and attractive flavours and colours to the meat.

In salting and curing, the main aim is preservation. This is achieved by high concentrations of salt which inhibit most micro-organisms. This can be achieved by either rubbing salt into the meat (salting) or by soaking in salt solution (curing or brining). Salting or curing of meat is practised in some countries to produce products such as salted pork or bacon.

Biltong is an example of a salted dried meat which is found in southern Africa. It consists of strips of meat which are dark brown in colour with a salty taste and a flexible rubbery texture. It is mainly used as a snack and has a shelf-life of several months under correct storage conditions.

It is produced by removing the fat and cutting the meat into thin strips. These are either soaked overnight in a mixture of salt and herbs/spices, or these ingredients are rubbed into the meat. It is then sun-dried by hanging the strips on a frame under an insect-proof mesh. If biltong is to be produced in more humid or cooler climates, a dryer could be used instead of sun-drying.

Production stages for biltong

Production stage	Equipment	Section reference
Wash meat	Clean water	
Trim the fat from the meat	Knife	
Slice	Cutting and slicing equipment	17.1 and 17.2
Rub salt and flavourings into the slices	Weighing equipment	64.1
Dry	Mats, racks Solar dryer Fuel-fired dryer Combined dryer	 23.1 23.2 23.3
Pack	Traditional packaging materials Heat sealing machines	 47.1

Biltong production

Cutting

To cut the meat into thin slices a sharp sterilized knife can be used. For more precise slicing, electrically-powered knives are available.

Salting

There are many methods of salting the product. Usually a mixture of salt and spices is prepared (the type and amount depend on individual tastes). This is then rubbed onto the pieces of meat in order to achieve a uniform distribution. Alternatively the meat is soaked in a solution of these ingredients.

Meat and meat products

Packaging

Salted, dried meat needs to be protected from moisture pick-up and attack by insects. If the climate is dry it may be packaged in traditional materials such as jute bags and cane baskets. For more humid areas, sealed polythene bags can be used.

Ground meat products

Here, the process of grinding meat into small pieces (such as minced meat) or pastes (sausage-meat), aims to change the texture of the meat and allow it to be formed into different shapes. Most commonly, these are cylindrical sausages enclosed by a casing (skin), or flat discs (patties). Both enable more rapid cooking, and both methods allow for spices and other flavours to be included.

The process of grinding the meat does not help to preserve it, and in fact causes more rapid spoilage. This is because there is more chance for bacteria from hands or dirty equipment to become mixed with the meat. Grinding also releases enzymes from the meat which cause changes to its flavour and texture.

Therefore, if sausage or patty-making is to be considered, the operators must be thoroughly trained in hygienic processing, all equipment should be thoroughly sterilized, and all processing should be done quickly, preferably at a low temperature (below 10°C) using ice or refrigeration, to slow down bacterial growth.

Preservation is achieved after grinding by one or more methods including adding chemical preservatives, refrigeration or freezing, smoking, drying or, pasteurization. It can be seen from these factors that ground-meat products have the potential to cause food poisoning and it is strongly recommended that productions are not attempted unless the staff are fully aware of the potential risks involved, and are trained to avoid them.

The table below outlines the production stages for sausages and patties.

Chopping/mincing

The cheapest option is to use a knife, but for a more thorough and efficient chopping process, bowl-choppers are available. They are designed specifically for chopping meat to be processed into sausages. They consist of a rotating, horizontal bowl and a set of rotating vertical knives. The chopper will transform the meat into a homogenous sausage emulsion.

Manual or powered mincers/grinders may be used to produce minced meat for burger patties. They consist of a screw shaft inside a metal barrel, and a cutter/die assembly at the

Equipment required

Processing stage	Equipment	Section reference
Washing		
Mince/chop	Cutting, slicing and dicing equipment	17.1, 17.2 and 17.3
	Bowl chopper	12.2
	Mincers and shredders	42.0
Add ingredients	Weighing and measuring equipment	64.1 and 64.2
Form/mould	Burger press	53.3
Extrude	Extruding machines (cold)	27.2
Casing (skin)	Filler	28.2 28.3
Pack	Heat sealing machine	47.1

end of the barrel. As the handle is turned, the screw rotates and the minced meat is pushed along the barrel, chopped by the knives, and forced out through the die.

Forming/moulding

Burger patties can be moulded by hand. The mixture is weighed into portions of the correct size, and moulded and pressed by hand. Whilst this method is effective, it also gives opportunity for bacteria from workers to contaminate the patty. Alternatively, burger and patty moulds are available quite cheaply or they can be easily fabricated from local materials.

Extrusion

Sausages are prepared by forcing the ground-meat mixture into casings. Extrusion can be achieved by using a hand-held plunger such as a piston type, which may be fitted with a range of different sized nozzles.

Packaging

These products are mostly sold uncooked. If not purchased at the site of production, they should be kept under refrigeration. In terms of packaging they require a covering to prevent contamination by insects and dirt. Refrigeration slows down the actions of enzymes and micro-organisms to give a shelf-life of 1–7 days. In the raw state there is a high risk of bacterial infection, and the need for refrigeration requires expertise and strict control over hygiene. These products may also be frozen to give the product a longer shelf-life of up to six months.

Production stages for sausages and patties

Process product	Wash meat	Mince/chop	Add ingredients	Form/mould	Extrude	Fill into skins	Pack
Patties	*	*	*	*			*
Sausages	*	*	*		*	*	*

Suitability for small-scale production

The potential hazards of infection, food poisoning and product deterioration make meat processing largely unsuitable for small-scale food processors. Only people with knowledge and experience of these hazards during routine production should consider becoming meat processors.

11. Fish and fish products

The small-scale fisheries of developing countries are vital because they provide a nutritious food which is often cheaper than meat and therefore available to a larger number of people.

In many countries the bulk of fish is sold fresh for local consumption. Processing, where this is done, is either to supply distant markets or to produce a range of products with different flavours and textures.

Fish is an extremely perishable food. For example, most fish become inedible within twelve hours at tropical temperatures. Spoilage begins as soon as the fish dies, and processing should therefore be done quickly to prevent the growth of spoilage bacteria. Fish is a low acid food and is therefore very susceptible to the growth of food poisoning bacteria. This is another reason why it should be processed quickly. Some methods of preservation cause changes to the flavour and texture of the fish which result in a range of different products. These include:

○ Cooking (for example, boiling or frying)

○ Lowering the moisture content (by salting, smoking and drying – collectively known as curing)

○ Lowering the pH (by fermentation)

Lowering the temperature with the use of ice or refrigeration also preserves the fish, but causes no noticeable changes to the texture and flavour.

The nutritional significance in the diet

Fish provides a good source of high quality protein and contains many vitamins and minerals. It may be classed as either white, oily or shellfish. White fish, such as haddock and seer, contain very little fat (usually less than 1%) whereas oily fish, such as sardines, contain between 10–25%. The latter, as a result of its high fat content, contain a range of fat-soluble vitamins (A, D, E and K) and essential fatty acids, all of which are vital for the healthy functioning of the body. The table opposite illustrates some of the main nutritional differences between oily and white fish.

Bringing in a catch of fish

Average composition of fish

Composition	White fish e.g. haddock	Oily fish e.g. herring
Energy (KJ)	321	970
Protein (g)	17	17
Fat (g)	0.7	18
Water (g)	82	64
Calcium (mg)	16	33
Iron (mg)	0.3	0.8
Vitamin A (μg)	0	45
Thiamine (mg)	0.07	0

Certain processing techniques such as boiling leach the water-soluble vitamins into the surrounding liquid. If this is thrown away, a great deal of nutritional value is lost.

Types of fish products

Cooked fish

Cooking provides short-term preservation of fish and it is usually a few days before any deterioration becomes noticeable.

A range of methods are used for cooking fish but the principle of the process remains the same. The flesh of the fish softens, enzymes become inactivated and the process kills many of the bacteria present on the surface of the fish.

Boiling and poaching both involve cooking the fish in hot water whereas frying uses hot oil. The advantage of these techniques is they are very simple and require no more than basic household equipment and are therefore suitable for small-scale production.

Cooked fish products are most usually for immediate consumption and require no sophisticated packaging. The shelf-life can be extended for a few days by using refrigerated storage and the product should be covered to prevent recontamination.

Cooled/frozen fish

The spoilage of fish is directly related to temperature. The higher the temperature, the faster the spoilage up to around 40°C, above which heat will destroy bacteria and enzymes. Any reduction in the temperature prior to processing will maintain the quality of the fish for longer.

Fish can be kept cool by covering it with clean, damp sacking and placing it in the shade. Although this method is simple and requires no special equipment, the fish still begins to deteriorate within a few hours.

An alternative is to pack the fish with ice. This is an effective method and preserves the fish for a longer period of time. Obtaining ice, however, can be difficult for the following reasons:

○ Most ice-making machines are power-operated and therefore require some kind of fuel. Obtaining fuel can often be difficult and the machines may prove expensive to operate.

○ A great deal of ice is required and often the cost of the ice is greater than the actual cost of the fish.

Freezing is an alternative method for cooling fish. This technique provides long-term preservation, but it is relatively expensive in terms of equipment and operating costs. In view of this it is not recommended for the majority of small-scale fisheries.

Cured fish products

Curing involves the techniques of drying, dry salting/brining (soaking in salt solution) or smoking. These may be used alone or in various combinations to produce a range of products with a long shelf-life. For example:

○ Drying – Smoking – Drying
○ Brining – Smoking – Drying
○ Salting – Drying
○ Salting – Drying – Smoking

Techniques such as these reduce the water content in the flesh of the fish, and thereby prevent the growth of spoilage micro-organisms.

Dried fish

The heat of the sun and movement of air remove moisture which causes the fish to dry. In order to prevent spoilage, the moisture content needs to be reduced to 25 per cent or less. The percentage will depend on the oiliness of the fish and whether it has been salted.

Traditionally, whole small fish or split large fish are spread in the sun on the ground, or on mats, nets, roofs, or on raised racks. Sun-drying does not allow very much control over drying times, and it also exposes the fish to attack by insects or vermin and allows contamination by sand and dirt. Such techniques are totally dependent upon the weather conditions. The ideal is dry weather with low humidity and clear skies.

Drying fish on racks

Fish and fish products

Figure 1 illustrates a solar tent dryer. This was first developed in Bangladesh, but there are now numerous variations in different parts of the world. It is probably one of the most simple designs.

Figure 2 shows an improved solar dryer, with a separate collector and drying chamber. The chimney is painted black to absorb more heat. This will heat the air inside the chimney, thereby increasing the air flow through the dryer.

Alternatives to sun-drying involve the use of solar or artificial dryers. There has been a great deal of research on the development of solar dryers as an improved method of drying fish. This has shown that by achieving increased drying temperatures and reduced humidities, solar dryers can increase drying rates and produce a lower moisture content in the final products, with improvements in fish quality compared with the traditional sun-drying techniques.

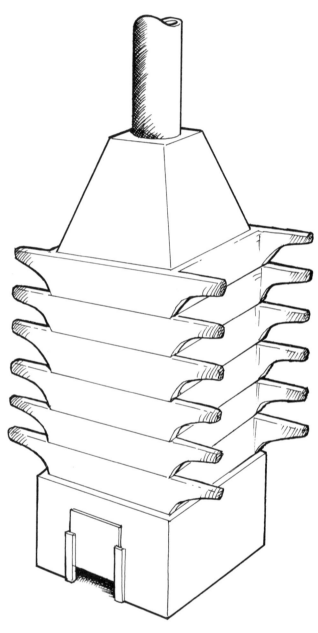

Figure 3: Tray dryer

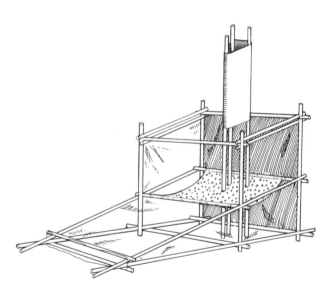

Figure 1: Improved solar dryer

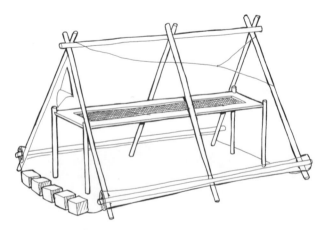

Figure 2: Solar tent dryer

Figure 3 illustrates an artificially-heated tray dryer. When rain threatens, the trays, which were previously placed in the sun to dry, are assembled on top of each other over a simple heating compartment. A roof and chimney are placed on top, and drying continues by direct heating.

Both solar and artificial dryers try to overcome the difficulties posed by sun-drying during the rainy season. With these dryers it is possible to minimize drying times and to increase the product quality. It should however, be pointed out that it is only advantageous to use such dryers if there is a market for a higher-quality product or if the fish would otherwise be lost.

Salted fish

Most food poisoning bacteria cannot live in salty conditions and a concentration of 6–10 per cent salt in the fish tissue will prevent their activity. The product is preserved by salting and will have a longer shelf-life. However, a group of micro-organisms known as 'halophilic bacteria' are salt-loving and will spoil the salted fish even at a concentration of 6–10 per cent. Further removal of the water by drying is needed to inhibit these bacteria.

During salting or brining two processes take place simultaneously:

○ water moves from the fish into the solution outside

○ salt moves from the solution outside into the flesh of the fish.

Salting requires minimal equipment, but the method used is important. Salt can be applied in many different ways. Traditional methods involve rubbing salt into the flesh of the fish or making alternate layers of fish and salt (recommended levels of salt usage are 30–40 per cent of the prepared weight of the fish). There is often the problem, however, that the concentration of salt in the flesh is not sufficient to preserve the fish, as it has not been uniformly applied. A better technique is brining. This involves immersing the fish into a pre-prepared solution of salt (36 per cent salt). The advantage is that the salt concentration can be more easily controlled, and salt penetration is more uniform. Brining is usually used in conjunction with drying.

Ultimately the effectiveness of salting for preservation depends upon:

○ uniform salt concentration in the fish flesh

○ concentration of salt, and time taken for salting

○ whether or not salting is combined with other preservation methods such as drying.

Smoked fish

The preservative effect of the smoking process is due to drying and the deposition in the fish flesh of the natural chemicals of wood smoke. Smoke from the burning wood contains a number of compounds which inhibit bacteria. Heat from the fire causes drying, and if the temperature is high enough, the flesh becomes cooked. Both of these factors prevent bacterial growth and enzyme activity which may cause spoilage.

Fish can be smoked in a variety of ways, but as a general principle, the longer it is smoked, the longer its shelf-life will be.

Smoking can be categorized as:

○ *Cold smoking.* In this method, the temperature is not high enough to cook the fish. It is not usually higher than 35°C.

○ *Hot smoking.* In this method, the temperature is high enough to cook fish.

Hot smoking is often the preferred method. This is because the process requires less control than cold processing and the shelf-life of the hot-smoked product is longer, because the fish is smoked until dry. Hot smoking does, however, have the disadvantage that it consumes more fuel than the cold-smoking method.

Traditionally, the fish would be placed with smouldering grasses or wood. Alternatively, fish may be laid or hung on bamboo racks in the smoke of a fire (see below).

Smoking fish traditionally

There are various types of kiln available in different parts of the world, which are used for smoking. Although traditional kilns and ovens have low capital costs, they commonly have an ineffective air-flow system, which results in poor economy of fuelwood and lack of control over temperature and smoke density. Improved smokers include the oil drum smoker and the chorker smoker.

Fish and fish products

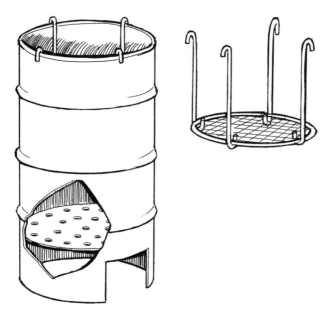

Oil drum smoker

Chorker smoker

As well as improved smokers, there are also improved techniques which involve either pre-salting the fish, so that the moisture content is reduced prior to smoking. Alternatively there are a range of improved kiln and oven designs (for details refer to the equipment catalogue section).

Production of cured fish products

The table below outlines the stages in the production of a range of products:

Equipment required

Processing stage	Equipment	Section reference
Gut	Cutting equipment	17.1 17.2
Wash		
Salt	Weighing and measuring equipment Brine meter	64.1 and 64.2 64.6
Smoke	Smoking equipment	60.0
Dry	Solar dryer Fuel-fired dryer Electric dryer	23.1 23.2 23.3
Pack	Packaging materials Sealing machine	47.1

Packaging of cured fish products

The most important concerns regarding packaging for these products are to prevent moisture pick-up and to prevent re-contamination by insects and micro-organisms. Traditional packaging materials include cane baskets, leaves, and jute bags. Alternatives include flexible packaging such as polythene bags, or wooden and cardboard packs. Indeed, the two may be combined, as in a polythene bag enclosed in an outer cardboard pack.

Fermented fish

Fermentation is a process by which beneficial bacteria are encouraged to grow. These bacteria increase the acidity of the fish and therefore prevent the growth of spoilage and food-poisoning bacteria. Additionally, salt is used to prevent the action of spoilage bacteria and allow the fish enzymes and the beneficial acid-producing bacteria to soften (break down) the flesh. Fermentation is therefore the controlled action of the desirable micro-organisms in order to alter the flavour or texture of the fish and extend the shelf-life.

The use of fermentation as a low-cost method of fish preservation is commonly practised all over the world. There are many different types of fermented products and their nature depends largely on the extent of fermentation which has been allowed to take place. They can be categorized as:

○ fish which retains its original texture

○ pastes

○ liquids/sauces.

Process/product	Gut	Wash	Treat with salt	Dry	Smoke	Dry	Pack
Dried fish	*	*		*			*
Dried and salted fish	*	*	*	*			*
Dry-salted and smoked fish	*	*	*	*	*		*
Brined and smoked fish	*	*	* Brine solution	*	*	*	*

As with salting, there is little need for equipment other than pans and containing vessels, and the process may easily be carried out on a small scale.

The table below outlines stages in the production of a typical fermented-fish product.

Fish paste (bagoong)

This is a product from Eastern Asia. It is made from whole or ground fish, fish roe, or shellfish. It is reddish brown in colour, although this will depend on the raw materials used, and is slightly salty with a cheese-like odour.

Equipment required

Processing stage	Equipment	Section reference
Wash	Clean water	
Drain		
Gut	Cutting equipment	17.1 and 17.2
Add salt (approx 5 per cent)	Weighing and measuring equipment	64.1 and 64.2
Leave to ferment (for several months)	Fermentation bin	03.1
Add colouring (optional)		
Pack	Sealing machines	47.1

Packaging of fermented fish

There are almost as many traditional methods of packaging fermented fish as there are ways of making it – such as earthenware pots, oil cans, drums and glass bottles. In the past, the latter have been used because of their low cost, but nowadays, cheaper plastic containers tend to replace the traditional types. The most important function of packaging for fermented fish products is that the containers should be air-tight, helping to develop and maintain the airless conditions required for good fermentation and storage. As the major advantage of these products is their low cost, the type of packaging is necessarily restricted. Glass bottles are often used for the better-quality products, but earthenware pots and even plastic bags are used.

Suitability for small-scale production

Although traditional processing represents a low-cost option for many small-scale producers, there may be large losses in terms of wasted fish. Improved technologies are usually techniques that require little in the way of expensive equipment, but at the same time increase the quality and the efficiency of the process. Often all that is needed to improve the process and the quality of the final product is the provision of clean water, education and training facilities, simple equipment, or basic materials.

It is important, as with all food processing ventures, to ensure that there is a market for the processed fish. Unfortunately, in areas where fresh fish is a more desirable commodity, small-scale fish processors, with their less-preferred cured products, may face fierce competition from larger-scale processors who have access to refrigeration and transport facilities.

12. Packaging

The main aims of packaging are to keep the food in good condition until it is sold and consumed, and to encourage customers to purchase the product. Correct packaging is essential to achieve both these objectives. The importance of packaging can be summarized as follows.

○ If adequately packaged, the shelf-life of local surpluses of food may be extended, and this allows the food to be distributed to other areas. In doing so, consumers are given more choice in terms of food available, food resources can be more equitably distributed, and rural producers may be able to generate income from surplus produce.

○ Correct packaging prevents any wastage (such as leakage or deterioration) which may occur during transportation and distribution.

○ Good packaging and presentation encourages consumers to buy products.

Solutions to packaging problems differ from region to region. Variations are the result of factors such as economics, the availability or access to packaging materials, infrastructure, distribution systems, climatic conditions and consumer habits. In many parts of the world, foods are wrapped in re-used newsprint, animal skins, rushes, or reeds. These materials are normally used for foods which are consumed soon after purchase (e.g. snack foods and bakery goods) and which therefore need little protection, or for foods such as flour and sugar which are likely to be transferred into storage vessels in the home.

Foods with a longer expected shelf-life have different needs and may require more sophisticated packaging to protect them against air, light, moisture, and bacteria.

Functions of packaging

Packaging should provide the correct environmental conditions for food starting from the time food is packed through to its consumption. A good package should therefore perform the following functions:

○ it should provide a barrier against dirt and other contaminants thus keeping the product clean

○ it should prevent losses. For example, packages should be securely closed to prevent leakage

○ it should protect food against physical and chemical damage. For example the harmful effects of air, light, insects, and rodents. Each product will have its own needs

○ the package design should provide protection and convenience in handling and transport during distribution and marketing

○ it should help the customers to identify the food and instruct them how to use it correctly

○ it should persuade the consumer to purchase the food.

Constraints on adequate packaging

Inadequate packaging may be the result of:

○ a lack of knowledge of the materials and/or the requirements for packaging different foods. Each product has its own characteristics and packaging requirements vary

○ in many countries the choice of packaging materials may be limited. For those that are available, supplies are often situated in urban areas and this may cause problems for the rural producer in terms of transportation and often in negotiating with suppliers

○ packaging can represent a large part of the total cost of a processed food. This may be in part the result of the higher unit cost when small quantities are ordered for small-scale production.

Packaging materials

In many developing countries the most commonly used packaging materials include:

○ leaves
○ vegetable fibres
○ wood
○ papers, newsprint
○ earthenware
○ glass
○ plastics
○ metals

Leaves

Banana or plantain leaves are often used for wrapping certain types of food (e.g. steamed doughs and confectionery). Corn husk is used to wrap corn paste or unrefined block sugar, and cooked foods of all types are wrapped in leaves (e.g. vine leaves, bamboo leaves). They are an excellent solution for the problem of how to package products that are consumed soon after purchase as they are cheap and readily available. They do not however protect the food against moisture, oxygen, odours or micro-organisms, and are therefore not suitable for long-term storage.

Stuffing leaves with food

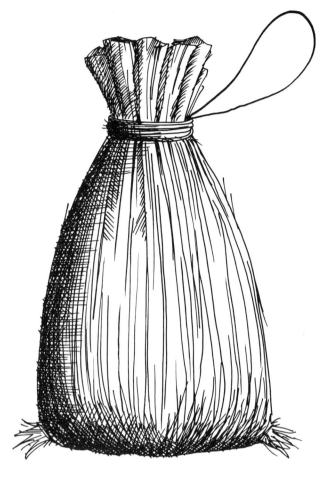

Peri mula

Vegetable fibres

These include bamboo, banana, coconut, cotton, jute, raffia, sisal, and yucca. These natural materials are converted into yarn, string or cord which will form the packaging material. These materials are very flexible, have some resistance to tearing, and are lightweight for handling and transportation. Being of vegetable origin, all of these materials are bio-degradable and to some extent re-usable.

As with leaves, vegetable fibres do not provide protection to food which has a long shelf-life since they offer no protection against moisture pick-up, micro-organisms, or insects and rodents.

Wood

Wooden shipping containers have traditionally been used for a wide range of solid and liquid foods including fruits, vegetables, tea and beer. Wood offers good protection, good stacking characteristics, strength and rigidity. Plastic containers, however, have a lower cost and have largely replaced wood in many applications. The use of wood continues for some wines and spirits because the transfer of flavour compounds from the wooden barrels improves the quality of the product.

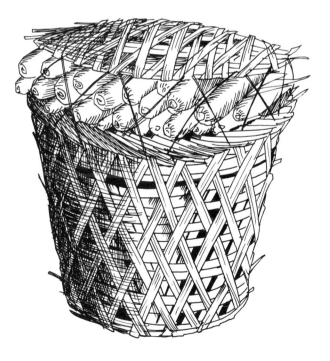

Vegetable fibre basket

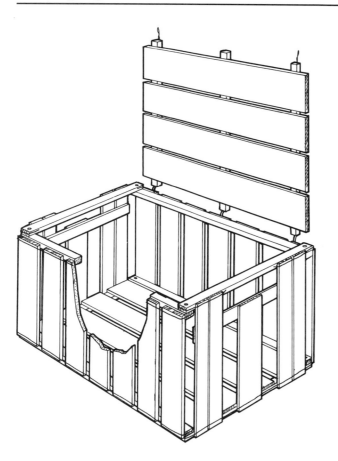

Wooden crate

Paper

Paper is an inexpensive packaging material and can be made from a wide range of materials, including rice husks, banana leaves and wood pulps. It is however highly absorptive, fairly easily torn, and offers no barrier to water or gases.

Some of these constraints can be overcome by treating the paper in various ways. A well-known method is to dip the paper in wax, or alternatively impregnate it with varnish or resin. Paper can also be strengthened by combining it with hessian cloth, cardboard or polythene.

The degree of paper re-use will depend on its former use, and therefore paper that is dirty or stained should be rejected. Newsprint should be used only as a outer wrapper and not be allowed to come into direct contact with food, as the ink used is toxic.

Earthenware

Earthenware pots are used worldwide for storing liquids and solid foods such as curd, yoghurt, beer, dried food, and honey. Corks, wooden lids, leaves, wax, plastic sheets, or combinations of these are used to seal the pots. Unglazed earthenware is porous and is suitable for products that need cooling such as curd. Glazed pots are needed for storing liquids (oils, wines) as they are light-proof, and if clean, restrict the entry and growth of micro-organisms, insects and rodents.

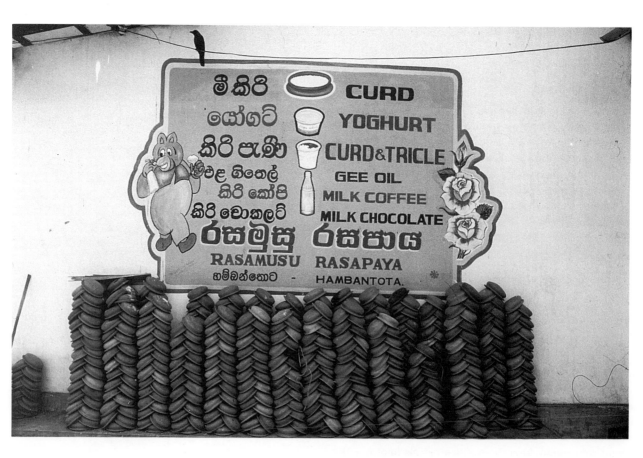

Curd pots

Glass

Glass has many properties which make it a popular choice as a packaging material:

A display of products in glass bottles/jars

○ glass is able to withstand heat treatments such as pasteurization and sterilization

○ it does not react with food

○ it is rigid and protects the food from crushing and bruising

○ it is impervious to moisture, gases, odours and micro-organisms

○ it is re-usable, re-sealable and recyclable

○ it is transparent, allowing products to be displayed. Coloured glass may be used either to protect the food from light or to attract customers.

Despite its many advantages, glass does have certain constraints:

○ glass is heavier than many other packaging materials and this may lead to higher transport costs

○ it is easy to fracture, scratch and break if heated or cooled too quickly

○ potentially serious hazards may arise from glass splinters or fragments in the food.

Preparation of glass containers

A good product packaged in a dirty container will soon deteriorate and therefore the following stages are recommended:

Inspection. This stage applies equally to both new and re-used containers. Any chipped, cracked, or heavily soiled containers must be rejected. Containers which have been used to store strong smelling substances such as kerosene should be rejected.

Washing. It is preferable that detergent and bleach are used for cleaning. However, if these are not available, soap or ashes may be used. Simple hand-held brushes can be used to aid the cleaning process on a small scale. For higher rates of production, powered washers are available (see Section 02.0).

Rinsing. Thorough rinsing of containers is essential. A simple rinsing system, illustrated below, involves a series of spigots set into a length of pipe which act as rinse sprays.

Bottle rinser

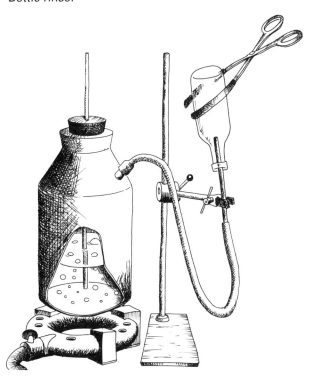

Sterilization of bottles

Sterilization. It is strongly recommended that prior to filling, glass containers are heat sterilized. Sterilization can be achieved very simply by holding the open neck of the container over the spout of a kettle containing boiling water. Containers can be inverted over a steam pipe which is supplied from a water boiler (see illustration). The vertical pipe is vital as it acts as a safety valve. Alternatively, glass containers can be boiled in water for ten minutes to sterilize them (see diagram below).

Sealing and capping

Bottles. Suitable machines for sealing bottles can be produced locally. There are two types of cap commonly used: those that thread onto the bottle neck, and crown caps which are pushed on under pressure. Simple low cost crown capping machines are available (see Section 47.2).

Jars. The sealer shown below is an efficient method for sealing jam jars with push-on lids (compared to domestic methods of using plastic and rubber bands). If bottle or jar necks are not always regular in size, a flexible plastic material allows the sealer to be fitted on different sizes of container. For details of other pieces of sealing equipment see section 47.1.

Cooling

Many products are filled into glass containers while they are hot and then need to be cooled quickly. To achieve this the containers can be stacked and cooled by circulating air. Alternatively, the cooler shown below can be used for both glass and metal containers.

A shallow bath is constructed and the newly filled containers enter at point A, and roll slowly down the sloping bottom to B, where they are removed. Cold water enters at B and overflows as hot water at A. A weak solution of bleach is allowed to drip into the end, B, thus ensuring that the cooling water is chlorinated.

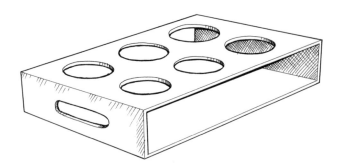

Bottle cooling system

A slightly different method is used for jars of jam. In order to set to a smooth gel, the jam jar must be allowed to cool standing in a vertical position. The filled jars are loaded into 'carriages' and then placed in the cooling trough, as for the bottles.

Plastics

The use of various plastics for containing and wrapping food depends on what is available in a particular country. Plastics are extremely useful as they can be made in either soft or hard forms, as sheets or containers, and with different thickness, light resistance, and flexibility. The filling and sealing of plastic containers is similar to glass containers.

Flexible films are the most common form of plastic. Generally, flexible films have the following properties:

○ their cost is relatively low

○ they have good barrier properties against moisture and gases

Jam jar sealer

○ they are heat sealable to prevent leakage of contents

○ they have wet and dry strength

○ they are easy to handle and convenient for the manufacturer, retailer, and consumer

○ they add little weight to the product

○ they fit closely to the shape of the product, thereby wasting little space during storage and distribution.

Flexible films include cellulose, polypropylene, and polythene.

Material	Properties
Cellulose	Plain cellulose is a glossy transparent film which is odourless and tasteless. It is tough and sharp resistant, although it tears easily. However it is not heat sealable and the dimensions and the permeability of the film vary with changes in humidity. It is used for foods that do not require a complete moisture or gas barrier.
Polythene Low-density polythene	Heat sealable, inert, odour-free, and shrinks when heated. It is a good moisture barrier, but has a relatively high gas permeability, sensitivity to oils, and poor odour resistance. It is less expensive than most films and is therefore widely used.
High-density polythene	Stronger, thicker, less flexible, and more brittle than low-density polythene. It has lower permeability to gases and moisture. It has a higher softening temperature (121°C) and can therefore be heat sterilized.
Polypropylene	Polypropylene is a clear glossy film with a high strength, and is puncture resistant. It has low permeability to moisture, gases, and odours, which is not affected by changes in humidity. It stretches, although less than polythene. It has good resistance to oil, and therefore can be used successfully for packaging oily products.
Coated films	Coated films are coated with other polymers or aluminium to improve the barrier properties or to enable them to be heat sealed.
Laminated films	Laminated films are two or more films glued together. Lamination improves the appearance, the barrier properties, or the mechanical strength of a package. Aluminium foil is widely used in laminated films where low gas, water vapour, odour, or light transmission is required.

Sealing with a candle and hacksaw blade

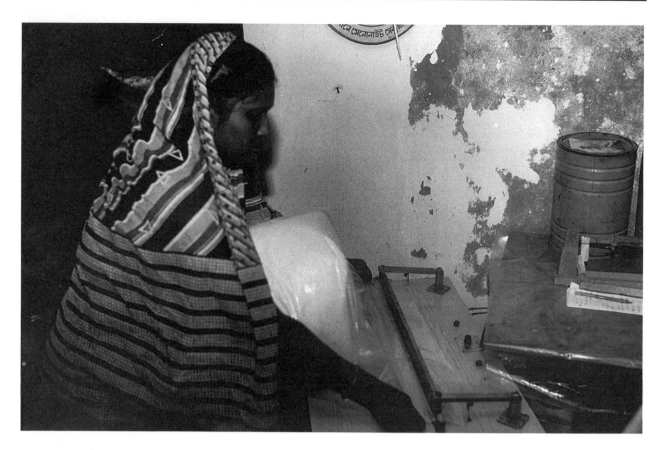

Sealing with a powered sealer

Sealing techniques

Sachets or pouches made from flexible films can be sealed by folding the edge of the film over the teeth of a hacksaw blade (a used hacksaw blade works better) and passing the folded edge through a flame. The result does not look as good as when a heat sealer is used, but practice can improve the appearance of the bag.

Powered sealers can be operated by placing the plastic between two heated bars (the bars are coated with a special tape to prevent the plastic from melting). These machines are available commercially and usually designed to use electricity (see section 47.1) but they can be reproduced locally and modified to use alternative energy sources.

It is possible on a small scale to seal foil lids onto plastic pots. This can be done by purchasing a specially-designed heat sealer (see section 47.1), or alternatively, a domestic iron can be used.

Metal

Metal cans have a number of advantages over other types of containers:

○ they provide total protection of the contents

○ they are tamperproof

○ they are convenient for presentation.

Constraints associated with use of metal cans include:

○ they are heavier than other materials, except glass, and therefore have higher transport costs

○ the heat treatment associated with the use of metal cans is not suitable for small-scale production (see sterilization).

Sealing techniques

Cans are sealed with a double seam and there are small scale seaming machines available to do this (see section 05.2).

The table opposite summarizes properties of a range of packaging materials.

Properties of packaging materials

Packaging materials such as glass are often made in developing countries but materials such as plastic film are more commonly imported from multinational packaging manufacturers. Most multinationals have a retail agent situated in developing countries and contact addresses can be found in local business directories.

Further information regarding suppliers and local costs of materials can be obtained from local packaging institutes; again these can be located through business directories.

Resistance to these factors

Type of packaging	Puncture crush etc	Sunlight	Air	Water	Heat	Odour	Insects	Rodents	Micro-organisms
Metal cans	*	*	*	*	#	*	*	*	*
Glass (bottle jar)	*	Coloured	*	*	*	*	*	*	*
Paper bag		*					#		
Cardboard	*	*			#		#		
Wood (box)	*	*			*		#	#	
Pottery (sealed lid)	*	*	*	*	*	*	*	*	*
Foil		*		#	#	*	#		
Plastic tub sealed	*	*		*	#	*	*	#	*
Cellulose uncoated			*	*		*	#		
Cellulose coated			*	*		*	*		*
Polyethylene Low-density			#	*		#	#		*
Stretch wrap				*					
Shrink wrap				*					
High density			*	*		*	#		*
Polypropylene			*	*		*	#		*
Polyester Plain			*	*		*	*		*
Metallized	*	*	*	*	*	*	#		*

* = good protection   # = some protection

PART TWO

DIRECTORY

01.0 Steam blanchers

Blanching is an important processing stage for a number of reasons: primarily it inactivates enzymes that cause deterioration in colour and flavour during drying and subsequent storage of some fruit and vegetables. In addition it improves the texture of the food and assists the preservation of vitamins. Blanching may be carried out using water or steam.

BLANCHER/SCALDER

This blancher is suitable for both fruits and vegetables. The material is placed in a basket and dipped into a bath of hot water. After a pre-set time, it is drained and placed in cold water for cooling. Scale is 60 × 60 × 45cm, with a capacity of 90-100kg.

PRICE CODE: 1

GARDENERS CORPORATION
6 Doctors Lane
PB No. 299
NEW DELHI 110 001
INDIA

TRAY-TYPE STEAM BLANCHER

The trays are placed in the blanching compartment, where the food is blanched by steam and then passed automatically to the cooling compartment. Scale is 1.8 × 0.8 × 1.2m.

PRICE CODE: 3

FMC CORPORATION
Machinery International Division
P.O. Box 1178
SAN JOSE
California 95108
USA

Lye peeling offers an effective way of peeling some fruits and vegetables. The food is placed in a hot solution (at or near boiling point) of sodium hydroxide for a specified time which varies, depending on the fruit or vegetable to be peeled. The loosened skin is removed by jets of water. Care is needed, as hot lye is very dangerous, and corrosive to some metal equipment.

Lye peeling is often combined with blanching in one complete operation.

BLANCHING & LYE PEELING EQUIPMENT

(See also page 131)
This consists of three stainless steel tanks. All trays included have perforated sliding lids and dropside handles.

PRICE CODE: 2

RAYLONS METAL WORKS
Kondivitta Lane
Andheri-Kurla Road
BOMBAY 400 059
INDIA

02.0 Bottle washing equipment

A variety of bottle washers are available. They include the following variations:
● simple manual bristle washer
● hydro washer (cleaning is performed by jets of water)
● soaker washer (bottles are completely immersed)
● combination of the above.

Manual bottle washers

BOTTLE BRUSHING UNIT

For cleaning bottles.

PRICE CODE: 1

CAN DRAINING

Consists of a rack for placing washed cans/pails/milk churns.

PRICE CODE: 1

MILK CAN WASHER

This is the steaming block type.

PRICE CODE: 1

DAIRY UDYOG
C-229A/230A, Ghatkopar Industrial Estate
L.B.S. Marg, Ghatkopar (W)
BOMBAY 400 086
INDIA

BOTTLE WASHER

This is suitable for bottles with a 9.5–10cm diameter. The unit consists of a spring-loaded nozzle which when depressed by an inverted bottle, produces jets of water. Throughput is 600 bottles per hour.

PRICE CODE: 1

ADELPHI MANUFACTURING CO. LTD
Olympus House
Mill Green Road
HAYWARDS HEATH RH16 1XQ
UK

BOTTLE WASHING MACHINE

This machine comprises a soaking tank made from galvanized iron. There is a 0.19kW (0.25hp) motor which turns brushes at 540rpm and a drain lock is provided for rinsing purposes.

PRICE CODE: 1

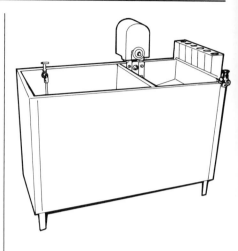

JT JAGTIANI
National House, Tulloch Road
Apollo Bunder
BOMBAY 400 039
INDIA

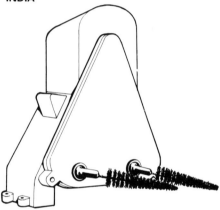

BOTTLE WASHER

This is a two-headed bottle washing machine driven by a 0.19kW (0.25hp) single-phase motor. It is supplied with four nylon brushes (two are spare) for washing inside bottles. A large brush is also provided for cleaning exteriors. Inside and outside are washed simultaneously. There are two tanks; one for hot water and one for cold.

PRICE CODE: 2

EMPTY CAN WASHER

This is a horizontal rotary-type washer for N-cans. The machine is capable of handling cans of sizes N 1-10. Cans are either washed by jets of water or sterilized by steam.

PRICE CODE: 2

RAYLONS METAL WORKS
Kondivitta Lane
Andheri-Kurla Road
BOMBAY 400 059
INDIA

Washers also manufactured by:

MASTER MECHANICAL WORKS PVT LTD
Pushpanjali SV Road
Santa Cruz (W)
BOMBAY 400 054
INDIA

NARANGS CORPORATION
25/90 Connaught Place
Below Madras Hotel
NEW DELHI 110 001
INDIA

Powered bottle washers

MEADOW BOTTLE WASHER

This is a small batch cleaning machine, consisting of a two compartment soap tank, brushing machine and a six-jet rinser set. The bottles are soaked in warm water or mild detergent. An overflow is fitted as there is a constant inflow of fresh water from the rinsing process. Requires: 220/240-1-50 AC. Capacity is 120 × ½-litre bottles.

PRICE CODE: 4

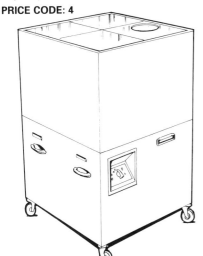

Bottle washers also manufactured by:

THE NORTHERN MEDIA SUPPLY LTD
Sainsbury Way
Hessle
HU13 9NX
UK

ADELPHI MANUFACTURING CO. LTD
Olympus House
Mill Green Road
HAYWARDS HEATH
RH16 1XQ
UK

03.0 Brewing equipment

03.1 Fermentation bins

RCT 127 CYLINDRICAL TANK

This simple cylindrical tank, complete with a fitted lid, is suitable for use as a fermentation bin. Capacity is 180 litres.

PRICE CODE: 2

White food grade polythene or polypropylene bins and tanks are also widely available in sizes from 10 to 200 litres.

MOULDED POLYTHENE CONTAINERS
Bridge Mills
Rochdale Road, Edenfield
RAMSBOTTOM
BL0 0RE
UK

03.2 Fermenters

AIR-LIFT FERMENTER SYSTEM

This system can be used for small-scale (including anaerobic) fermentations. Most organisms or cells can be grown and the equipment can easily be cleaned and maintained. In addition, there are no moving parts, bearings, etc. An electronic thermometer is supplied. Total volume: 4.5 litres.

PRICE CODE: 3

GB BIOTECHNOLOGY LTD
4 Beaconsfield Court
Sketty
SWANSEA
SA2 9JU
UK

04.0 Butter pats

Butter pats are of simple design and are used to knead and form the butter in order to remove excess moisture, produce a uniform texture, and make it into a form more convenient for use and storage.

LARGE WOODEN BUTTER PATS

These will knead up to 1kg of butter.

PRICE CODE: 1

BUTTER PATS

This is a set of two small wooden butter pats.

PRICE CODE: 1

PLASTIC TRAYS

A small tray used during the butter-kneading process.

PRICE CODE: 1

GEBR. RADEMAKER
P.O. Box 81
3640 AB MIJDECHT
NETHERLANDS

BUTTER SCOOPS

These are used for scooping and handling butter.

PRICE CODE: 1

DAIRY UDYOG
C - 229A/230A,
Ghatkopar Industrial Estate
L.B.S.Marg,
Ghatkopar (W)
BOMBAY 400 086
INDIA

Butter scoops are also manufactured by:

R J FULLWOOD & BLAND LTD
ELLESMERE
Shropshire
SY12 9DF
UK

05.0 Canning equipment

Perishable foods which are to be stabilized for extended storage must be sterilized to retain desirable quality factors (flavour, colour, aroma, etc.); to destroy enzymic activity; and most importantly to kill micro-organisms such as *Clostridium botulinum* which produce deadly toxins, and others which cause undesirable changes in the food quality. Of the different sterilization processes, canning is perhaps the most widely used. The process involves filling the food into a can, fitting the lid using a double seam and heating the can in a retort to sterilize the food. Different retort temperatures and processing times (derived from several complex equations and calculations) are used depending on the type of micro-organisms to be destroyed. Commercially, canning is carried out as either a 'batch' or 'continuous' process. For canning on a small scale, a batch system using retorts or pressure cookers is more appropriate.

Although canning is possible at a small scale, it is not a technology that is recommended, owing to several constraints. For example, the equipment, material and operating costs are high. As well as the retort and canning equipment, the process requires other equipment such as a steam generator and an air compressor. In addition, canning requires experienced workers to operate the equipment; maintenance engineers to service the equipment and microbiologists for examination of the products (thus also requiring access to microbiology media and equipment). It is a technology that should only be undertaken by experienced processors, especially if low acid foods such as fish, vegetables or meats are to be canned.

05.1 Retorts, sterilizers and pressure cookers

VERTICAL AUTOCLAVES

This autoclave consists of a brass inner, mild steel outer body together with a safety valve, pressure gauge and steam release cock. Working pressure is 138–152kPa (20–22psi), with a power range of 1.5–6kW.

PRICE CODE: 3

JT JAGTIANI
National House
Tulloch Road
Apollo Bunder
BOMBAY 400 039
INDIA

*CANNING RETORT

This retort is fitted with a dial thermometer, pressure gauge, safety valve, petcock, packing gland and winged clamps. Four models are available (see table below).

PRICE CODE: 3

RAYLONS METAL WORKS
Kondivitta Lane
Andheri-Kurla Road
BOMBAY 400 059
INDIA

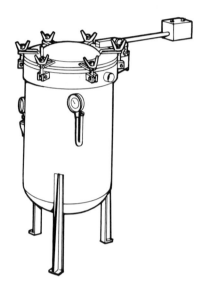

INSTRUMENT STERILIZER

This sterilizer is fitted with an immersion heater which has a safety device to cut off the current if the boiler becomes dry. It uses a 230V AC single-phase motor. Size range: 1–2kW.

PRICE CODE: 2

JT JAGTIANI
National House
Tulloch Road
Apollo Bunder
BOMBAY 400 039
INDIA

AUTOCLAVE

This has valves for purging, safety pressure relief and discharge of water and condensates. It is also fitted with a gauge and thermometers. Four models are available.

PRICE CODE: 2

DISENOS Y MARQUINARIA JER S.A.
Emiliano Zapata 51
Col Buenavista, 54710
Cuautitlan Itzcalli
MEXICO

PRESSURE COOKER/CANNER

For quick cooking or small-scale canning using a separate heat source. Both the body and the lid are cast from aluminium. The bottom is machined to ensure good surface-to-surface contact with the heat source and to improve heat transfer. Three sizes are available:

MODEL 15.5 Holds 10 pints (5.7litres)
MODEL 21.5 Holds 18 pints (10.2 litres)
MODEL 30   Holds 19 pints (10.8 litres)

PRICE CODE: 1

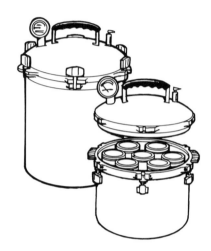

GIANT CANNER

This canner can be used on wooden stoves and hotplates. The top edge is wrapped around a 0.64cm bar of solid stainless steel. Silver is soldered and sandwich-locked seams are approved for direct contact with food. Two sizes are available: one with a capacity of 54.6 litres liquid and the other with a capacity of 31.8 litres.

PRICE CODE: 1

LEHMAN HARDWARE & APPLIANCES INC.
P.O. Box 41
4779 Kidron Road
Kidron
OHIO 44636
USA

Canning equipment also manufactured by:

GARDENERS CORPORATION
6 Doctors Lane
PB No. 299
NEW DELHI 110 001
INDIA

REPUTE SCIENTIFIC CO.
13/21 3rd Panjarapole Lane
2nd Floor, SVU Bg CP Tank Road
BOMBAY 400 004
INDIA

NARANGS CORPORATION
25/90 Connaught Place
Below Madras Hotel
NEW DEHLI 110 001
INDIA

BABAR STAINLESS STEEL IND (PVT) LTD
Siegt Rd
GUJRANWALA
PAKISTAN

*CANNING RETORT MODEL	CAN TYPE (figures represent no. processed per batch)			
	A10	A2.5	LLB TALL	5.5 OZ
Diam 610 × H 760mm	28	115	225	600
Diam 610 × H 915mm	35	135	260	750
Diam 680 × H 915mm	75	225	400	1000
Diam 810 × H 915mm	95	240	360	1230

ARMFIELD TECHNICAL EDUCATION
CO. LTD
Bridge House
West Street
RINGWOOD
BH24 1DY
UK

FMC CORPORATION
Machinery International Division
P.O. Box 1178
SAN JOSE
California 95108
USA

RR MILL INC.
45 West First North
Smithfield
UTAH 84335
USA

05.2 Seamers

CANNING MACHINE

This is a small machine with an automatic shut-off system and a 0.19kW (0.25hp) motor. It is equipped with serrated plates for 552cc and 858cc cans, and seals 240–360 cans per hour. Requires 125V, 60Hz.

PRICE CODE: 2

INDUSTRIAL DE PARTES S.A. DE C.V.
Latoneros Num. 99
Col Trabajadores del Hierro
MEXICO DF 02650
MEXICO

BENCH-TYPE DOUBLE SEAMER

This has a 0.19kW (0.25hp) motor. Cans are seamed by placing the can and cover on the base plate, and raising the base plate which then clamps the can between the chuck and plate. When the lever is pulled the automatic seaming operation starts. Throughput is 200–300 cans per hour; size is 6.8cm diameter × 10.2cm height.

PRICE CODE: 2

FMC CORPORATION
Machinery International Division
P.O. Box 1178
SAN JOSE
California
95108
USA

05.3 Miscellaneous canning equipment

HOT LIFTING TONGS
Used for lifting hot cans of all sizes.

PRICE CODE: 1

NARANGS CORPORATION
25/90 Connaught Place
Below Madras Hotel
NEW DEHLI 110 001
INDIA

Also manufactured by:

RR MILL INC.
45 West First North
Smithfield
UTAH 84335
USA

06.0 Carbonating equipment

Manual carbonators

IRA SODA WATER MACHINE

This machine makes carbonated drinks using high-pressure carbon dioxide. A pressure regulating valve supplies CO_2 from a gas cylinder at an even pressure. A filler/crowner is operated by holding the bottle and pulling a lever. The bottle is sealed by pressing a lever. Capacity is 600 bottles per hour. A motorized version is also available.

PRICE CODE: 2

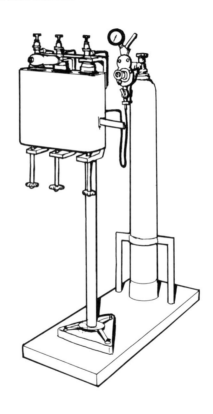

ESSENCE & BOTTLE SUPPLY (INDIA) LTD
14 Radha Bazar street
CALCUTTA 700 001
INDIA

Powered carbonators

MIRA BACHELOR SODA WATER MACHINE

This is a powered version of the above model. Filled bottles are placed in a cage and rotated. The gas and water are mixed and bottles are then capped using a hand operated crown capper. Throughput is 36, 72, or 108 bottles per hour.

PRICE CODE: 2

ESSENCE & BOTTLE SUPPLY (INDIA) LTD
14 Radha Bazar street
CALCUTTA 700 001
INDIA

07.0 Centrifuges

07.1 Dairy centrifuges

Dairy centrifuges are used to separate the cream from liquid milk, by the application of centrifugal force. The simplest type of equipment is the tubular bowl centrifuge, comprising a vertical cylinder or bowl which rotates inside a stationary casing at between 1500rpm and 5000rpm depending on the diameter. The milk is introduced continuously at the base of the bowl and the two liquids (cream and skimmed milk) are separated into layers which emerge from the two outlets. A variety of disc-bowl centrifuges are also available. In these a stack of conical discs inside a rotating bowl are used to separate the cream.

HAND-OPERATED MILK TEST CENTRIFUGE
This consists of a table and centrifuge.

PRICE CODE: 2

DUDYOG HAND-OPERATED MILK TEST CENTRIFUGE

PRICE CODE: 2

DUDYOG MILK TEST CENTRIFUGE MACHINE

This electric centrifuge has a speed indicator.

PRICE CODE: 2

DUDYOG SWISS MODEL

This electric centrifuge can carry out 36 tests for milk.

PRICE CODE: 2

DAIRY UDYOG
C-229A/230A, Ghatkopar Industrial Estate
L.B.S. Marg, Ghatkopar (W)
BOMBAY 400 086
INDIA

BEL FRESH CREAM

Makes small quantities of cream, with a capacity of 0.3 litres.

PRICE CODE: 1

MINI-CREM 80

Throughput is 80 litres per hour.

PRICE CODE: 1

RJ FULLWOOD & BLAND LTD
ELLESMERE
Shropshire
SY12 9DF
UK

Dairy and honey centrifuges

BATCH-TYPE THREE-POINT CENTRIFUGE

This centrifuge is driven by a single- or two-speed vertical motor running at 1440 rpm. All models require a 3-phase 50Hz AC power supply. Capacity is 25, 35, 60kg.

PRICE CODE: 2

K S SEETHARAMAIAH & SONS PVT LTD
29/1 Jaraganahalli
10th K M Kanakapura Road
BANGALORE 560 078
INDIA

ELECREM ELECTRIC CREAM SEPARATOR

Four models are available.

MODEL	CAPACITY (litres)
80	80
1	125
2	200
3	315

PRICE CODE: 3

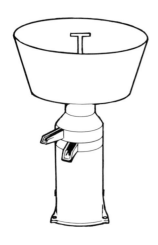

ELECREM 5

This machine separates cream on a larger scale. It is available with a 50 litre receiving vessel or a 140 litre mounted wall-tank, and requires 110V. Throughput is 500 litres per hour.

PRICE CODE: 4

R J FULLWOOD & BLAND LTD
ELLESMERE
Shropshire
SY12 9DF
UK

CREAM SEPARATOR

This model comprises a motor, barrel, skimmed milk catcher, cream catcher, inlet jar and whole milk reservoir. The thickness of the cream can be regulated. Throughput is 80 litres per hour.

PRICE CODE: 2

Other versions are available with throughputs of 125–500 litres per hour.

GEBR. RADEMAKER
P.O. Box 81
3640 AB MIJDECHT
NETHERLANDS

Dairy centrifuges also manufactured by:

LEHMAN HARDWARE & APPLIANCES INC.
P.O. Box 41
4779 Kidron Road
Kidron
OHIO 44636
USA

07.2 Honey centrifuges

Honey centrifuges operate by forcing honey out of the combs by centrifugal force. The extractor comprises a cylindrical container with a centrally-mounted fitting to support combs or frames of uncapped honey. There is also a mechanism to spin the frames which throws the honey out of the uncapped combs to the inner wall of the bowl. The honey then collects in the bottom of the bowl. A honey gate allows the honey to be drained out when required. Electrically operated high speed extractors increase the speed gradually to prevent damage to the combs.

Manual honey centrifuges

Country	Manufacturer	Equipment name	Capacity (kg)	Price code	Other information
Australia	Guilfoyle, John L (Sales) Ltd	2-frame bench s. steel extractor		1	H: 57cm Diam: 43cm
		2-frame bench extractor		1	H: 62cm Diam: 54cm Other sizes are available
		4-frame non-reversible extractor	45	1	H: 88cm Diam: 60cm
		2-frame reversible extractor	45	1	H: 88cm Diam: 60cm
	Pender Beekeeping Supplies Pty Ltd	Budget bench extractor		1	
		Economy bench extractor		1	
		2-frame manual reversible extractor		1	Produces 91kg/h. Other sizes are available
		Bench extractor		1	Produces 36kg/h
		4-frame manual reversible extractor		2	
		4-frame non-reversible bench extractor		2	Optional manual/ electric power
Austria	Puff, Stefan GmbH	Honey extractor model mini 3		1	
Denmark	Swienty	8001 3-frame extractor		2	Other sizes are available
		8080 4-frame extractor		2	Other sizes are available
USA	Dadant & Sons Inc.	Junior bench extractor	79	2	Throughput is 2 frames/load
		Little wonder extractor	136	2	Throughput is 4 frames/load

Powered honey centrifuges

Country	Manufacturer	Equipment name	Price Code
Australia	Guilfoyle, John L. (Sales) Ltd	2-frame bench steel extractor motor	2
		Electric motor	2
	Pender Beekeeping Supplies Pty Ltd	Auto self-radial honey extractor	2
		Auto self-centring base drive	2
Austria	Puff, Stefan GmbH	Reversible honey extractor	2

07.3 Juice centrifuges

DYNAMIC PAG1

This is a citrus fruit juice extractor. Halved fruits are placed in the top of the extractor and hand pressure on the lever activates the motor. Maximum juice is extracted whilst the pips and pulp are retained in the extractor. Throughput is one × 170g glass of juice in 30 seconds, and the machine's weight is 6kg.

PRICE CODE: 2

MITCHELL & COOPER LTD
Framfield Road
UCKFIELD
TN22 5AO2
UK

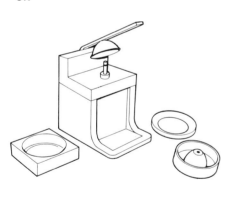

CENTRIFUGE JUICER

This is suitable for apples, pineapples etc. The machine has a 0.37kW (0.5hp) motor.

PRICE CODE: 2

NARANGS CORPORATION
25/90 Connaught Place
Below Madras Hotel
NEW DEHLI 110 001
INDIA

07.4 Filter centrifuges

The centrifuge illustrated is an example of a manual filter centrifuge.

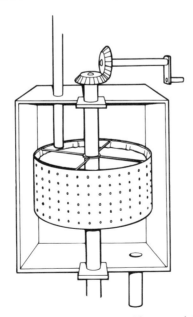

This can be made by taking a stainless steel drum with perforated holes (the drum of an old washing machine is perfect) and attaching it to a central shaft.

The shaft is made to rotate through the use of a toothed cogs and a flywheel.

The drum is encased in an outer container. This can be made from any material but if the filtrate is to be collected, the container must be made from food grade plastic or a similar material. At the top of the outer container there is an inlet feed chute for pouring the liquid and at the base there is a collecting chute.

MODE OF OPERATION

The liquid is fed in at the top through the inlet chute. Simultaneously, the handle is turned and the drum is made to rotate. This rotation causes the material to be thrown against the drum wall and the filtrate drains out through the perforated holes and is finally collected at the base. If necessary the drum can be lined with a filter cloth.

08.0 Cheese cloths

ALL PRICE CODE: 1
Manufactured by:

RJ FULLWOOD & BLAND LTD
ELLESMERE
Shropshire
SY12 9DF
UK

GEBR. RADEMAKER
P.O. Box 81
3640 AB MIJDECHT
NETHERLANDS

LEHMAN HARDWARE & APPLIANCES INC.
P.O. Box 41
4779 Kidron Road
Kidron
OHIO 44636
USA

09.0 Cheese moulds/ presses and kits

09.1 Cheese moulds

PLASTIC CHEESE MOULDS

These moulds can be used for making small cheeses. They comprise a series of plastic cylindrical moulds with followers. Capacity: 454g. Dimensions: 8 × 10cm.

PRICE CODE: 1

COULOMMIER MOULD

This is a mould for cheese production, comprising two stainless steel hoops of diameters 15 and 11.5cm.

PRICE CODE: 1

Other dimensions are also available.

CHEESE MOULD

This is a cylindrical stainless steel cheese mould with two hardwood followers. Two sizes are available: 18 × 13cm, and 15 × 11.5cm.

PRICE CODE: 1

RJ FULLWOOD & BLAND LTD
ELLESMERE
Shropshire
SY12 9DF
UK

CHEESE MOULDS

These wooden moulds have stainless steel hoops which are used for holding and moulding the cheese. Six sizes are available: 0.5, 1, 2, 3, 3.5 and 4kg.

PRICE CODE: 1

CHEESE MOULD

This is an earthenware mould complete with a cheese cloth. Capacity: 500g.

PRICE CODE: 1

In addition, various plastic cheese moulds are available without cloths and with capacities of 650g or 2kg.

CHEESE MOULD

This is a plastic cheese mould without cloths. Capacities: 650g or 2kg.

PRICE CODE: 1

CHEESE MOULDS

These special moulds are for soft cheese. Different moulds are available for: Ricotta, Camembert, Roquefort, Brie, and Baby Gouda cheeses. Draining stones are also available.

PRICE CODE: 1

GEBR. RADEMAKER
P.O. Box 81
3640 AB MIJDECHT
NETHERLANDS

09.2 Cheese presses

CHEESE PRESS

This is a table-top model, which can also be used for pressing fruit. The press comes supplied with a mould, drip-tray, weights and recipes. It is capable of making 1.4–1.8kg cheeses.

PRICE CODE: 1

R&G WHEELER
Hoppins
Dunchideock
EXETER
EX2 9UL
UK

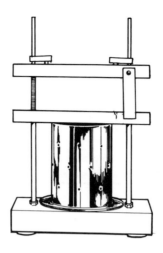

DELUXE CHEESE PRESS

This machine does not require any weights. Hoops are available for pressing 9 litres or 18 litres. Recipes and instructions are included.

PRICE CODE: 1

A low-cost cheese-press kit is also available (**PRICE CODE: 1**).

LEHMAN HARDWARE & APPLIANCES INC.
P.O. Box 41
4779 Kidron Road
Kidron
OHIO 44636
USA

Many other versions are available from:

GEBR. RADEMAKER
P.O. Box 81
3640 AB MIJDECHT
NETHERLANDS

R J FULLWOOD & BLAND LTD
ELLESMERE
Shropshire
SY12 9DF
UK

09.3 Cheese kits

CHEESE BOARD

This is a small board for holding cheeses up to 10cm².

PRICE CODE: 1

SOFT CHEESE KIT

This small-scale soft cheese package comprises: muslin, starter culture, rennet, stainless steel ladle, and recipe book.

PRICE CODE: 1

COULOMMIER CHEESE KIT

This is a complete kit for small-scale cheese production. It comprises a stainless steel mould, cheese mats, two cheese boards, rennet, starter culture, and a recipe.

PRICE CODE: 1

RJ FULLWOOD & BLAND LTD
ELLESMERE
Shropshire
SY12 9DF
UK

CHEESE KIT

This is a complete kit for making cheese, comprising a press, two earthenware cheese moulds with cheese cloths, one large cheese cloth for curd, one curd-cutter, two thermometers, one bottle of calf rennet, one cheese culture , one yoghurt culture and one recipe book. The kit makes two 500g cheeses.

PRICE CODE: 1

GEBR. RADEMAKER
P.O. Box 81
3640 AB MIJDECHT
NETHERLANDS

10.0 Cheese vats

CHEESE VAT

This machine is equipped with a variable-speed stirrer, and a chart recorder measures the pH of the milk. Water is circulated through the jacket at a pre-selected temperature. Capacity is 1kg of cheese from a 10 litre batch of milk.

PRICE CODE: 2

ARMFIELD TECHNICAL EDUCATION CO. LTD
Bridge House
West Street
RINGWOOD
BH24 1DY
UK

11.0 Chocolate conches

Refining and conching are two important steps in the manufacture of chocolate. Refining involves the crushing and shearing of cocoa and sugar particles to produce a product with the desired particle size. Chocolate for the UK or USA markets should be refined to an average particle size of between 20μm and 30μm, whereas European tastes demand a lower size range of 15μm to 22μm, which gives a smoother product.

Conching is used to develop flavour and to remove undesirable volatile components (e.g. acetic acid), produced during the fermentation of the cocoa beans. The duration of conching is an important factor because most of the flavour changes occur within the first 24 hours. Likewise the temperature at which conching takes place is important, in order to achieve the desired taste/texture. For example, milk chocolate should be conched at 45° to 60°C, whereas plain chocolate should be conched at 55° to 85°C for British taste and up to 100°C for European tastes.

Generally the refining/conching process is as follows. Cocoa powder, icing sugar and the required fats are placed in the feed hopper of the unit. Heating and blending takes place and air is drawn through the mixture. The mix is circulated through a unit in which rotating cylinders press against spring loaded grinding bars to achieve the required particle size reduction. The chocolate mass is then held at the required temperature for a predetermined time to conche it. It should be noted that high quality chocolate production is difficult to achieve technically (requiring experience and expertise) and that the equipment is generally expensive. However, re-melting commercial chocolate for coating/re-using is a viable small-scale industry in many countries.

CHOCOLATE REFINER/CONCHE

This machine can be supplied with an automatic tensioning device to give the most efficient use of the motor. Capacity: 50–100kg.

PRICE CODE: 4

BCH EQUIPMENT
Mellor Street
ROCHDALE
OL12 6BA
UK

ACMC TABLE-TOP TEMPERER

This machine can be used for melting and tempering chocolate. It comprises a 5.7 litre stainless steel bowl, a polyethylene scraper and a thermometer holder. The heat source is a 2.1kW lamp operating on 110V AC (also available for 230V AC).

PRICE CODE: 3

GFE BARTLETT & SON LTD
Maylands Avenue
HEMEL HEMPSTEAD
HP2 7EN
UK

CHOCOMILL CHD250

This is a refiner for chocolate pastes. The mixture is fed into the grinding chamber by means of a heated pump mounted on the outside of the machine. It is then subjected to high shear by the steel discs operating at speed. A screen at the front of the machine allows finished chocolate to pass through. Capacity: 18 litres. Throughput: 200kg per hour.

PRICE CODE: 4

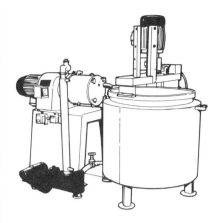

ALUMCRAFT ARGENTINA S.C.A.
Mercedes 568
1407 BUENOS AIRES
ARGENTINA

 # 12.0 Choppers

12.1 Fruit and vegetable choppers

HERB MILL

This mill is used for chopping small amounts of fresh herbs.

PRICE CODE: 1

LAKELAND PLASTICS
38 Alexandra Buildings
WINDERMERE
LA23 1BQ
UK

BEAN CHOPPER

This diagonally cuts string beans. The unit houses three blades and will clamp to any surface between 0.6 and 2.5cm thick. Height: 17.8cm.

PRICE CODE: 1

LEHMAN HARDWARE & APPLIANCES INC.
P.O. Box 41
4779 Kidron Road
Kidron
OHIO 44636
USA

ROTATING CHOPPER

Used primarily for household- or small-scale operations, this chopper is suitable for solid or leafy vegetables. The degree of fineness can be controlled.

PRICE CODE: 1

RR MILL INC.
45 West First North
Smithfield
UTAH 84335
USA

TOMATO CHOPPER

This chops tomatoes into small pieces suitable for canning or further processing. The unit comprises a 0.75kW (1hp) three-phase motor, and a circular ring of 35.6cm diameter with holes of 2.5cm. It also has eight beaters which pulverize the tomatoes. A larger version is available with a throughput of 1.25 tonnes per hour.

PRICE CODE: 2

GARLIC CHOPPER

The chopping blades are made from high carbon steel and are copper plated. The knives and the bowl turn simultaneously. A baffle turns the garlic pieces for finer chopping. The machine uses a 1.4kW (2hp) motor. Throughput is 600–900kg per hour. Scale (bowl): Height 25cm, Diameter 76cm.

PRICE CODE: 3

RAYLONS METAL WORKS
Kondivitta Lane
Andheri-Kurla Road
BOMBAY 400 059
INDIA

12.2 Bowl choppers

BOWL CHOPPER SM45 HS

This chopper is used for the manufacture of all types of sausage. A digital display indicates bowl speed and mixing temperature. The hand starter determines the bowl speed and the speed of the knife shaft. Standard knives will cut, mix and emulsify the meat into an homogeneous sausage emulsion. The power of the motor is 6.8kW (9hp). Dimensions: length 138 × width 95 × height 115 cm.

PRICE CODE: 4
KRAEMER UND GREBE GMBH & CO.
Postfach 2149 & 2160
D-3560 Biedenkopf
WALLALL
GERMANY

MEDIUM AND LARGE MEAT GRINDERS/CHOPPERS

These can be used for medium scale meat chopping. They are similar to the small models except that they do not clamp but are bolted to the surface for more stability. Size range: Models 12, 22, 32. A small meat grinder is also available.
PRICE CODE: 1

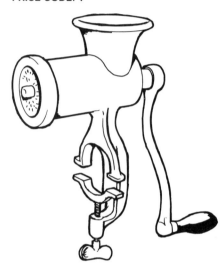

MEAT CHOPPER ACCESSORIES

Three meat chopper attachments are available: stuffers for sausage filling; belt drive pulleys for replacing manual operation; and plates for making hamburgers, sausages and lard.
PRICE CODE: 1
LEHMAN HARDWARE & APPLIANCES INC.
P.O. Box 41
4779 Kidron Road
Kidron
OHIO 44636
USA

Meat choppers also manufactured by:
TALSABELL S.A.
Poligono Industrial V
Salud 8-46.950
VALENCIA
SPAIN

 # 13.0 Butter churns

13.1 Manual churns

HAND-OPERATED BUTTER CHURN

This churn is used for small-scale butter production and comprises a glass jar with a handle which turns the churning blade. Capacity is 3 litres. A version with a small motor attached is available (*Manuelba 20 Butter Churn*).
PRICE CODE: 1
R J FULLWOOD & BLAND LTD
ELLESMERE
Shropshire SY12 9DF
UK

BUTTER CHURN

This wooden butter churn is mounted on a cast iron stand, and comes equipped with a clamping device. Sizes range from 20 litres to 125 litres.
PRICE CODE: 1
LAKSHMI MILK TESTING MACHINERY CO.
A90 Group Industrial Area
Wazirupa
NEW DELHI 110 052
INDIA

BUTTER CHURN

Three models are available: cedar, redwood or shatterproof plastic.
PRICE CODE: 1
LEHMAN HARDWARE & APPLIANCES INC.
P.O. Box 41
4779 Kidron Road
Kidron, OHIO 44636
USA

Butter churns also manufactured by:
DAIRY UDYOG
C-229A/230A, Ghatkopar Ind. Est.
L.B.S. Marg, Ghatkopar (W)
BOMBAY 400 086
INDIA

GEBR. RADEMAKER
P.O. Box 81
3640 AB MIJDECHT
NETHERLANDS

J.J.BLOW LTD
Oldfield Works
Chatsworth Road
CHESTERFIELD S40 2DJ
UK

13.2 Butter churns (powered)

ELECTRIC BUTTER CHURN

This churn comprises a plastic jar and electric motor, with variable speed control. Churning time is between 15 and 25 minutes. Capacity: 5 or 10 litres. A stainless steel version is also available.
PRICE CODE: 2

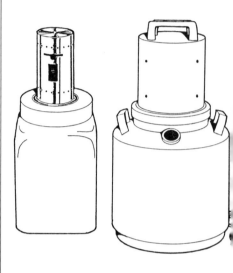

BUTTER CHURN/KNEADER

This machine washes and kneads butter in one process. It comprises a wooden tub and 'dasher', and the dasher's blades work the cream. It is driven by an electric motor (380V) with slow or fast speeds. A sight-glass measures the liquid level, and a tap drains the buttermilk. Capacity is 100–160 litres.
PRICE CODE: 2
GEBR. RADEMAKER
P.O. Box 81
3640 AB MIJDECHT
NETHERLANDS

Powered butter churns also manufactured by:
RJ FULLWOOD & BLAND LTD
ELLESMERE
Shropshire
SY12 9DF
UK

 # 14.0 Cleaners

14.1 Fruit and vegetable cleaners

WASHING MACHINE

This machine is used to wash fruit and vegetables and is fabricated from carbon steel. Two models are available: one that immerses the product, and another using a rotating perforated drum. Both versions have a

0.75kW (1hp) motor and a cylinder diameter of 91cm. Dimensions: 380 × 152 × 170cm. Throughput: two tonnes per hour.

PRICE CODE: 4

MAPISA INTERNATIONAL S.A. DE C.V.
Eje 5 Oriente Rojo Gomez 424
Col Agricola Oriental
MEXICO DF 08500
MEXICO

ROOT CROP WASHER

This machine consists of a peeling drum with an abrasive surface, a hopper and a mesh slide. A water spray is provided throughout the whole operation in order to facilitate the process. A suitable batch size is 15–20kg per hour, with an overall throughput of 100kg per hour.

PRICE CODE: 2

VISAYAS STATE COLLEGE OF
AGRICULTURE
8 Lourdes Street
PASAY CITY 3129
PHILLIPINES

LV 2000 VEGETABLE WASHER

This model gently washes salads, fruits and fish. It comes complete with a centrifugal drier for salads. The machine will also defrost fish. Capacity: leafy vegetables 3kg, root crops 15kg. Dimensions: 91 × 60 × 60 cm.

PRICE CODE: 4

CRYPTO PEERLESS LTD
Bordesley Green Road
BIRMINGHAM B9 4UA
UK

WASHER/PEELER, MODEL ROT 3

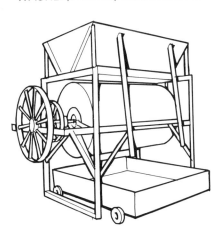

This is a batch rotary washer/peeler, designed for cassava and equipped with a 3.7Kw (5hp) electric motor. The machine comprises a wooden cylinder with a water spray. Throughput: 100–160kg gari/h. Dimensions: 120 × 150cm. Other sizes are also available.

PRICE CODE: 3

INDUSTRIAS MAQUINA D'ANDREA S.A.
Rue General Jardim 645
SAO PAULO
01223
BRAZIL

POTATO WASHER

This machine requires a 2.7kW (3.6hp) motor. Throughput: 200kg per hour.

PRICE CODE: 4

INDUSTRIAS TECHNICAS 'DOLORIER'
Alfredo Mendiola 690
Urbanizacion Ingenieria
LIMA
PERU

14.2 Grain cleaners

Grain cleaners are designed for use after winnowing primarily to separate all chaff, straw, weed seeds, broken and inferior seeds, dust, and other contaminants.

Cleaning and grading the grain makes it more suitable for a number of uses: grain required for seed purposes should be clean and uniform; better prices can be obtained for graded samples; and removal of insects and other foreign matter during cleaning ensures better and safer storage. In addition, pre-cleaned grain improves the efficiency of the drying machine, during the drying process, and results in a more uniformly-dried sample. Some machines will simultaneously clean and grade a sample of grain or seed, while others will perform only one of these functions.

PORTABLE GRAIN CLEANER

This machine is constructed of wood and steel (an International Rice Research Institute design), and comprises a single shaft, horizontal oscillating screen and fan. It uses a 0.37kW (0.5hp) motor or alternatively a 0.75kW (1hp) petrol engine, and has a multi-crop capability owing to the use of two screens, the top one being interchangeable. The weight of the cleaner plus motor is 72kg. Length 123cm, width 67cm, height 126cm. A grain purity of up to 98% can be obtained. Fuel consumption is 0.5 litres per hour, and throughput is up to 1000kg per hour of rice.

PRICE CODE: 3

JCCE INDUSTRIES
242 Mayondon
Los Banos
Laguna
PHILIPPINES

A similarly-designed cleaner is manufactured by:

ALPHA MACHINERY & ENGINEERING CORP.
1167 Pasong Tamo Street
MCC Makati
METRO MANILA D 708
PHILIPPINES

GRAIN CLEANER

The machine consists of a hopper, sieve and grain outlet and is powered by a 0.5kW (0.7hp) single phase 1500rpm electric motor, with a throughput of 1000kg per hour of paddy.

PRICE CODE: 2

TANZANIA ENGINEERING &
MANUFACTURING DESIGN ORGANIZATION
P.O. Box 6111
Arusha
TANZANIA

SEED AND GRAIN CLEANER

This comprises an all-steel construction with pneumatic feeding, variable air regulation, two screens, and two air-separators. There is also an integral three-phase 3.7kW (5hp) motor. Length 316cm, width 146cm, height 256cm. Throughput: 700–1000kg per hour of wheat.

PRICE CODE: 2

ORIENTAL MACHINERY (1919) PVT LTD
25 R N Mukherjee Road
CALCUTTA 700 001
INDIA

PADDY CLEANER

This cleaner removes stones and dust from paddy. The dust blower uses an aspirator to remove light impurities and dust, while the lower part of the machine contains vibrating sieves to eliminate the larger impurities. Throughput is 2–3 tonnes per hour.

PRICE CODE: 3

DEVRAJ & COMPANY
Krishan Sudama Marg
Firozpur City
PUNJAB
INDIA

MOBILE VILLAGE SEED CLEANER

This cleaner is used for cereal grains. It has a tubular frame construction, and is mobile with a small petrol or diesel engine. There are screening slots of 2.11mm, 2.24mm or 4.37mm, and lipped riddles of 9.5mm or 12.7mm apertures. The cleaner has a weight of 550kg, with a throughput of 1–1.5 tonnes per hour.

PRICE CODE: 3

INLAND SALES & SERVICE
11–13 Railway Parade
Merredin, WA 6415
AUSTRALIA

TWO-SCREEN GRAIN SEED-CLEANER/GRADER

Used primarily for wheat, this cleaner is able to clean both large and heavy grains as well as small and light seeds, due to its adjustable air separation. A three-screen version is also available. Length 140cm, width 102cm, height 147cm. Screen size: 62 × 57cm. Motor size 0.75kW (1hp). Throughput: 300–600kg per hour.

PRICE CODE: 2

COSSUL & CO. PVT LTD
123/367 Industrial Area
Fazalgunj
Kanpur
UTTAR PRADESH
INDIA

SEED CLEANER/GRADER

This comprises an all-metal construction with built-in dust collectors and pneumatic feed, with variable air regulation. In addition there are two screens, each of dimensions 143 × 112cm, and two air-separators. The machine requires a three-phase 5.6kW (7.5hp) electric motor. Overall size is length 306cm, width 250cm, height 257cm, and throughput is 1500–2000kg per hour (wheat).

PRICE CODE: 2

SHAKING-TYPE UNPOLISHED RICE SEPARATOR

This machine requires a 0.37kW (0.5hp) motor. Throughput is 850–1200kg per hour. Larger models are available.

PRICE CODE: 2

KOREA FARM MACHINE & TOOL INDUSTRIAL CO-OP.
11–11 Dongja-Dong
Youngsan-Gu
SEOUL
KOREA

Grain purifiers also manufactured by:

DEVRAJ & CO.
Krishan Sudama Marg
Firozpur City
PUNJAB
INDIA

Cleaners (wheat/paddy) also manufactured by:

DANDEKAR MACHINE WORKS LTD
Dandekarwadi, Bhiwandi
Dist. Thane
MAHARASHTRA 421 302
INDIA

Cleaners/graders also manufactured by:

BEHERE'S & UNION INDUSTRIAL WORKS
Jeevan Prakash
Masoli
DAHANU ROAD 401 602
Dist Thane
INDIA

CV KARYA HIDUP SENTOSA
JL. Magelang 144
YOGYAKARTA 55241
INDONESIA

15.0 Cold storage

ICELESS REFRIGERATOR

This cooled storage unit is not commercially available. The unit comprises an open timber frame cupboard with metal trays. Burlap is cut and fitted around the unit. Information and design can be obtained from:

ILO – SDSR
P.O. Box 60598
NAIROBI
KENYA

DOMESTIC GAS REFRIGERATOR

This is used for cooling and storing food. The unit has an electrical backup system and locking doors. It uses 8.6 litres of gas per week, and has a storage capacity of 16.8 litres. Total shelf space = 1.04m². In addition, there is extra storage space in the side of the door. Dimensions: Height 162 × width 63 × depth 63cm.

PRICE CODE: 4

KEROSENE REFRIGERATOR

The unit is filled with kerosene and lighted at the back. It can hold over 13.6 litres of kerosene which is enough for four weeks' operation. Designed for the tropics, the cooling unit is able to maintain full cooling and freezing capacities in temperatures up to 50°C. The fridge has a storage space of 0.8m² and the freezer compartment has a atorage space of 0.1m². Spare parts are readily available for the refrigerator.

PRICE CODE: 4

'BEST' GAS REFRIGERATOR

This can be used for freezing and cooling foodstuffs. Full freezing performance can be maintained in temperatures of over 38°C. There is an optional flip-switch for electrical backup cooling. The cabinet is rust-resistant and has door storage. It uses 3.2kg of gas per week. In total there is 0.9m² shelf space and 0.3m² space in the freezer.

PRICE CODE: 4

LEHMAN HARDWARE & APPLIANCES INC.
P.O. Box 41
4779 Kidron Road
Kidron
OHIO 44636
USA

Refrigerators also manufactured by:

T A INDUSTRIAL DIVISION
P.O. Box 3910
HARARE
ZIMBABWE

QUICKFREEZE (PVT) LTD
P.O. Box 368
Southerton
HARARE
ZIMBABWE

16.0 Curd making equipment

16.1 Curd cutters

A range of stainless steel and plastic cutters are manufactured by:

GEBR. RADEMAKER
P.O. Box 81
3640 AB MIJDECHT
NETHERLANDS

16.2 Curd tubs

These are used for making and storing curd. They are specially designed for curd, being resistant to whey corrosion (aluminium and several plastics are not suitable), and they also insulate the product. There is a range of six sizes available, from 15 to 230 litres. Also in this range are: one plastic tub (80 litres); one enamel cauldron (27 litres) which can be placed on a gas cooker; and an enamel cauldron (27 litres) which has a heating element and a thermostat.

PRICE CODE: 1 (LARGEST = PRICE CODE 2)

CURD-MAKING SET

This contains everything needed to make curd, including a special bag for draining the curd, a bottle of rennet, a cheese culture and a description of the process.

PRICE CODE: 1

GEBR. RADEMAKER
P.O. Box 81
3640 AB MIJDECHT
NETHERLANDS

17.0 Cutting, slicing and dicing equipment

17.1 Cutters

Manual cutters

ROTARY CUTTER

This is used for cutting biscuits etc., and comprises a 25cm cutting wheel.

PRICE CODE: 1

GARDENERS CORPORATION
6 Doctors Lane
PB No. 299
NEW DELHI 110 001
INDIA

CIRCLE COOKIE CUTTERS

These are made from stainless steel and can be used to cut plain or fluted shaped biscuits. Size range: Diameter 5, 6.5, 7.5, 9, 10, and 11cm.

PRICE CODE: 1

MINI BISCUIT CUTTERS

This set comprises 12 different shaped mini biscuit cutters. They are also suitable for canapes and aspic jelly.

PRICE CODE: 1

LAKELAND PLASTICS
38 Alexandra Buildings
WINDERMERE LA23 1BQ
UK

CHEESE CUTTER

This cutter has two handles to give better control when cutting cheeses.

PRICE CODE: 1

GEBR. RADEMAKER
P.O. Box 81
3640 AB MIJDECHT
NETHERLANDS

COREMASTER II

This cores and cuts fruit. The fruit is cored and simultaneously wedged leaving a neat 1.2cm diameter core which can easily be separated from the wedges. Available for 6 or 8 wedges per fruit. A WEDGEMASTER II version is also available which cuts fruit into 6, 8, 10, or 12 wedges.

PRICE CODE: 1

MITCHELL & COOPER LTD
Framfield Road
UCKFIELD TN22 5AO2
UK

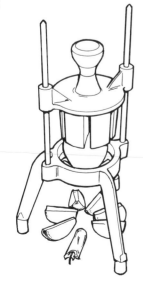

PINEAPPLE PUNCH

This hand operated machine cuts pineapples into round slices for canning.

PRICE CODE: 1

NARANGS CORPORATION
25/90 Connaught Place
Below Madras Hotel
NEW DELHI 110 001
INDIA

CROWN CUTTER

This hand-operated machine cuts fruit and vegetables into crown- or basket-shapes. There are three slicing rings which can cut various sizes of produce. This machine is ideal for preparing tomatoes, oranges, melons and grapefruit.

PRICE CODE: 2

MITCHELL & COOPER LTD
Framfield Road
UCKFIELD TN22 5AO2
UK

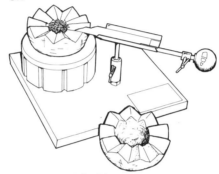

TOMATO PRO 1V

This tomato slicer has a pusher-head which forces the fruit through a fixed set of blades. A choice of blade spacing is available and maintenance simply consists of a blade tension adjustment and occasional replacement. Dimensions: length 50.8 × width 25.4 × height 25.4cm. Other models available are ECONOPRO, and TOMATO PRO II.

PRICE CODE: 2

APPLE CORER AND DIVIDER

A simple metal tool that removes the core and cuts each apple into fourteen equal segments.

PRICE CODE: 1

LOBSTER KING SAFETY CUTTER

This safety opener cuts fresh or frozen lobsters and clams. With a guide adjusted for a centreline cut, the stainless steel blade splits the shell so that the flesh can be easily removed in one piece. When fitted with the 1.3cm chopping board on the platform, the Lobster King can also cut pineapple, cabbage, root vegetables and cheese. Dimensions: 50.8 × 30.5cm.

A MINI LOBSTER KING version is also available.

PRICE CODE: 2

All the above from:
MITCHELL & COOPER LTD
Framfield Road
UCKFIELD TN22 5AO2
UK

TATER KING SCOOPER

This hand-operated scoop skins and wedges potatoes easily and quickly. Cooked potatoes are halved then scooped to obtain four uniform wedges per half for deep frying or roasting.

PRICE CODE: 1

MITCHELL & COOPER LTD
Framfield Road
UCKFIELD TN22 5AO2
UK

ROOT CROP STRIP-CUTTER

This cutter may be used for cutting root crops into french-fry sticks. It is provided with two sets of horizontal-cutting blade assemblies to cut peeled root crops. The machine is driven by a pedal mechanism utilizing the chain and pedals of a normal bicycle. Throughput is 86kg per hour.

PRICE CODE: 2

VISAYAS STATE COLLEGE OF
AGRICULTURE
8 Lourdes St
PASAY CITY 3129
PHILIPPINES

CABBAGE CUTTER

This cutter slices cabbage and other firm fruit and vegetables, using serrated stainless steel knife blades. The vegetable is placed in the wooden tray and the knife box is pushed over it. Length 53 × width 20cm.

PRICE CODE: 1

FRENCH-FRY CUTTER

This cuts potatoes into thin french-fry chips. Two sizes are possible: regular or shoestring. The potato is placed inside the machine and the handle is pressed downwards. The machine has a rust resistant, tin plated body with a plastic base. Throughput is 25 regular or 49 shoestring per cut. Length 22.9 × width 7.6cm. A spiral cut french-fry cutter is also available.

PRICE CODE: 1

LEHMAN HARDWARE & APPLIANCES INC.
P.O. Box 41
4779 Kidron Road
Kidron
OHIO 44636
USA

BREAD DIVIDER

This cuts dough into small pieces or balls prior to baking. Dough is placed in a circular steel ring under cutting blades. The top is pressed down and the dough is cut to form round balls of equal shape and size.

PRICE CODE: 1

KLAUY NAM THAI THOW OP
1505 07 Rama 4 Rd, Wangmai
Patumwan
BANGKOK 10330
THAILAND

Powered cutters

KOMET CRUSHER

This cutting machine can reduce the size of nuts, kernels or copra. It works like a pair of scissors, cutting the material without loss of oil, which would otherwise occur through milling, chopping, tearing or squeezing. The machine can also be used to crush the residue after the first pressing. Power requirements: 220/380V, 150Hz, 1.1kW (1.5hp) 1410rpm, (V-belt drive). The unit can be driven by petrol/diesel or from a tractor output shaft. Throughput is 100/300kg per hour. Dimensions: length 100 × width 40 × height 50cm.

PRICE CODE: 3

IBG MONFORTS + REINERS GMBH & CO.
Postfach 20 08 53
Monchengladbach 2
D-4050
GERMANY

CHIPPER RCII

This machine cuts potatoes into chips, with a choice of ten sizes. The standard equipment includes one universal rotor and one knife block, together with a built-in hopper and safety switch. The unit is powered by a 0.25kW (0.3hp) motor, and throughput is 1500kg per hour. Dimensions: 53 × 36 × 68cm.

PRICE CODE: 4

CRYPTO PEERLESS LTD
Bordesley Green Road
BIRMINGHAM B9 4UA
UK

LIME QUARTERING MACHINE

This can be used to quarter and prepare limes. The machine has four circular knives, driven by four 0.19kW (0.25hp) motors, which rotate at 650rpm in the same direction towards the centre. The feeder is available in the following sizes; 3.8, 5.1, 6.4, 7.6cm diameter, to accommodate limes of different sizes. Throughput: 1016kg sour limes per shift.

PRICE CODE: 3

RAYLONS METAL WORKS
Kondivitta Lane
Andheri-Kurla Road
BOMBAY 400 059
INDIA

R2 CUTTER, SLICER, JUICER

This model has three different attachments: the cutter attachment is suitable for chopping, mixing and blending operations; the slicer attachments are available with eight different discs; and the juicer has two different sized cones. The machine has a 0.19kW (0.25hp) single-phase motor and

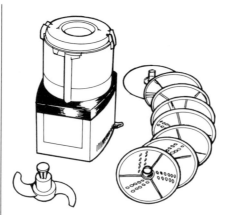

requires a 110/220V, 50Hz electricity supply. Bowl capacity: 2 litres. Dimensions: 39.6 × 21.3 × 36.5cm. An additional machine, MODEL 17, can be used for cutting, slicing, shredding, and grating.

PRICE CODE: 2

SPANGENBERG BV
Grootkeuken Professionals
De Limiet 26
4124 PG VIANEN
NETHERLANDS

DOUGH CUTTER

This machine cuts dough into 18–36 portions, each weighing 20–100g. Dimensions: 50 × 60 cm.

PRICE CODE: 2

MAQUINARIA OVERENA S.A. DE C.V.
Av. Division del Norte 2894
Col Parque San Andres
MEXICO DF 04040
MEXICO

CUTTING, SLICING, SHREDDING, MODEL 17

This machine can cut, slice, shred, and grate (using different discs).

PRICE CODE: 3

WODSCHOW & CO. A/S
Industrisvinget 6
Postbox 10
2605 BROENDBY
DENMARK

FOOD CUTTER

This can be used for cutting meat and vegetables. Two models are available: FC14 and FC19. The FC14 has a 0.37kW (0.5hp) motor with a capacity of 2–3kg, whilst the FC19 has a 0.75kW (1hp) motor and a capacity of 6–7kg. Both have trimming boards and an aluminium body.

PRICE CODE: 4

CRYPTO PEERLESS LTD
Bordesley Green Road
BIRMINGHAM B9 4UA
UK

Steam cutters

PLAIN UNCAPPING KNIVES

These remove wax layers from uncapped honey cells. They are the bent-handle type with a blade length of either 25 or 30cm. The knives should be periodically placed in boiling water to aid uncapping.

PRICE CODE: 1

PUFF, STEFAN GMBH
Neuholdaugasse 36
8011 Graz
AUSTRIA

PLAIN 254MM HONEY KNIFE

This is used for uncapping small amounts of honey comb. Weight 0.91kg.

PRICE CODE: 1

DADANT & SONS INC.
Hamilton
IL 62341
USA

UNCAPPING KNIVE

This knive is steam-heated and has a 25cm blade.

PRICE CODE: 1

STUART ECROYD BEE SUPPLIES
P.O. Box 5056
Papanui
Christchurch 5
NEW ZEALAND

Steam-heated knives also manufactured by:

PENDER BEEKEEPING SUPPLIES PTY. LTD
17 Gardiner Street
Rutherford
NSW 2320
Australia

B.J. ENGINEERING
Swallow Ridge
Hatfield
NORTON
Worcester WR5 2PZ
UK

HONNINGCENTRALEN A/L
Ostensjoveien 19
0661 Oslo 6
NORWAY

17.2 Slicers

SHREDDER/SLICER

This can be used for fruit and vegetables. It comprises a series of stainless steel blades, a plastic body and a 'finger-safe' food pusher. The slices can be wafer thin, medium or thick, and the shreds can be one of two thicknesses.

PRICE CODE: 1

EGG SLICER

This slices eggs in the conventional way, with the advantage of being hand-held.

PRICE CODE: 1

THREE-DRUM FOOD MILL

This hand-operated grater, slicer and shredder has a stainless steel rotary drum for each operation. It can be used right- or left-handed.

PRICE CODE: 1

**LAKELAND PLASTICS
38 Alexandra buildings
WINDERMERE
LA23 1BQ
UK**

VEGETABLE SLICER VS-2

This is used for slicing soft and semi-soft vegetables. It can cut up to 35 slices, 0.32cm thick. A safety shield is included which can be removed for cleaning. Weight: 10kg.

PRICE CODE: 2

ONION KING II

This slicer is used for slicing onions and other firm produce. The produce is positioned under the puller head, and with one swift pull, the whole item is sliced. Variations on this model are available.

PRICE CODE: 2

**MITCHELL & COOPER LTD
Framfield Road
UCKFIELD
TN22 5AO2
UK**

FRUIT AND VEGETABLE SLICER

This slicer can be used for most fruits and vegetables. The thickness of slices can be adjusted.

PRICE CODE: 1

**NARANGS CORPORATION
25/90 Connaught Place
Below Madras Hotel
NEW DELHI 110 001
INDIA**

ELLIOT HAND-OPERATED SLICER

This slices pineapple into a hollow cylinder shape for canning. The cored pineapple fruits are fed by hand to a V-block, located underneath the cutting assembly. When the pineapple fruit rests on the knife assembly, it passes through a series of slots which cut the pineapple cylinder into slices. Throughput: 600–720 pineapple fruits per hour. Dimensions: 91 × 61 × 168cm.

PRICE CODE: 3

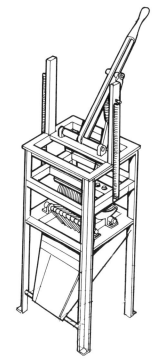

**FMC CORPORATION
Machinery International Division
P.O. Box 1178
SAN JOSE
California 95108
USA**

BEAN SLICER

This slicer cuts beans from end to end into strips. Beans are fed into the hopper and the handle is turned. Three blades are housed in the slicer, and as the beans are sliced they become tenderized, improving the flavour. The machine will clamp to any surface between 3.8 and 3.5cm. Height: 9.5cm.

PRICE CODE: 1

**LEHMAN HARDWARE & APPLIANCES INC.
P.O. Box 41
4779 Kidron Road
Kidron
OHIO 44636
USA**

POTATO CHIPPER

This potato chipper can be mounted on a bench or on a pedestal. Chipping is done by means of a hand-operated lever that presses the potato. The chipper is available in either a heavy- or light-duty version.

PRICE CODE: 1

**MITCHELL AND JOHNSON PVT LTD
P.O. Box 966
BULAWAYO
ZIMBABWE**

Manual slicing equipment also manufactured by:

**GARDENERS CORPORATION
6 Doctors Lane
PB No. 299
NEW DELHI 110 001
INDIA**

R 301 ULTRA PROCESSOR

This multi-purpose preparation machine slices, dices, shreds, grates, kneads and emulsifies. The cutting and mixing tools are optional extras. It uses a 220V, 50Hz single-phase electricity supply.

PRICE CODE: 2

**SPANGENBERG BV
Grootkeuken Professionals
De Limiet 26
4124 PG VIANEN
NETHERLANDS**

GRAVITY-FEED SLICERS

This range of meat slicers comprises three models, each with different blade size. GS22 has a 20cm blade, the GS25 has a 25cm blade and the GS30 has a 30cm blade. All models have a slice adjustment of 0–0.15cm, built-in knife sharpeners, and are belt-driven with a manual-feed carriage.

PRICE CODE: 3

**CRYPTO PEERLESS LTD
Bordesley Green Road
BIRMINGHAM
B9 4UA
UK**

BREAD SLICING MACHINE

A 36-blade arrangement, powered by a 0.37kW (0.5hp) motor, slices loaves up to 33cm in length. Throughput is 450 loaves per hour. Size range: Mini, medium, heavy.

PRICE CODE: 2

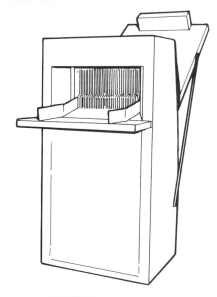

**BIJOY ENGINEERS
Mini Industrial Estate
P.O. Arimpur, Trichur District
KERALA 680 620
INDIA**

Bread slicers also manufactured by:

**SUPREMA EQUIP PARA IND DE PANIFICA
Estrada Municipal SMR 340
N 532/600 Jardim Boa Vista
SUMARE SP
BRAZIL**

**KLAUY NAM THAI THOW OP
1505 07 Rama 4 Rd, Wangmai
Patumwan
BANGKOK 10330
THAILAND**

VEGETABLE SLICING MACHINE

This machine includes a 0.37kW (0.5hp) motor and reduction gear. The cutting knives are mounted on metal holders and rotate in a cast aluminium hopper. Vegetables are fed into the hopper, sliced, and then fall onto a removable stainless steel tray. Width 46 × Depth 69cm.

PRICE CODE: 3

THAKAR EQUIPMENT COMPANY
66 Okhla Industrial Estate
NEW DELHI 110 020
INDIA

HALLDE RG 100

This general-purpose slicer can be used for slicing, dicing, shredding, chopping and grating. It has a wide feed opening to minimize the need for pre-cutting. Long products such as cucumbers can be fed through a large feed tube for continuous cutting. There is a 0.25kW (0.3hp) motor which uses a 240V electricity supply. Cutting tools (made from stainless steel) for slicing, dicing, or grating are also available. Throughput: 300kg vegetables per hour. Dimensions: 42.5 × 22.3 × 59.5cm.

PRICE CODE: 3

SPANGENBERG BV
Grootkeuken Professionals
De Limiet 26
4124 PG VIANEN
NETHERLANDS

R 301 ULTRA PROCESSOR

This multi-purpose preparation machine slices, dices, shreds, grates, kneads and emulsifies. The cutting and mixing tools are optional extras. It uses a 220V, 50Hz single phase electricity supply.

PRICE CODE: 2

ROBOT COUPE
10 Rue Charles Delescuze
BP 135
BAGNOLET 93170
FRANCE

17.3 Dicing equipment

ROOT CROP CUBER/SORTER

This machine uses pre-sliced roots at 1.25cm thickness as the raw material for cubing. The pre-slicing is done using a simple hand-operated cutter. Throughput: 150kg per hour.

PRICE CODE: 1

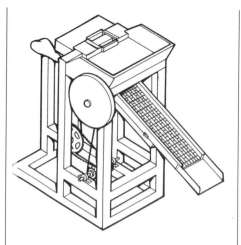

VISAYAS STATE COLLEGE OF
AGRICULTURE
8 Lourdes St
PASAY CITY 3129
PHILIPPINES

RAW PAP'AYA CUBING MACHINE

This machine dices papaya for further processing. The machine is fitted with a die and a punch to give 0.6cm cubes. It uses two rollers, one acting as the cubing roller and the other as a plain type. It is fitted with a 0.75hp (0.56kW) three-phase motor and gear reduction box. Other versions available: Pedal (50kg per hour), power operated (125kg per hour).

PRICE CODE: 2

RAYLONS METAL WORKS
Kondivitta Lane
Andheri-Kurla Road
BOMBAY 400 059
INDIA

 # 18.0 Decorating equipment

CREAM WHIPPER/DECORATOR

This hand-operated machine whips and pipes cream. It comes complete with five nozzle attachments for piping and garnishing. Total capacity: 0.25 litres whipped cream.

PRICE CODE: 1

MULTI-COLOUR ICING GUN

This machine comprises a dispensing head with separate reservoirs for different coloured icing sugars. There is also a lightweight icing gun for single-handed operation. The set contains six different sized nozzles.

PRICE CODE: 1

BISCUIT STAMPS

This set adds decorative detail to baked products. The small hand-held ceramic stamps leave an impression when pressed onto a shortbread or biscuit mixture. Four designs are available: rose, thistle, celtic cross, or dragon.

PRICE CODE: 1

LAKELAND PLASTICS
38 Alexandra Buildings
WINDERMERE LA23 1BQ
UK

Lakeland Plastics also manufacture the following equipment:
COMPLETE BISCUIT MAKER
TALA FLOWERS
ICING SETS

 # 19.0 Decorticators

Manual decorticators

UNATA 5200

This hand-operated combined decorticator and winnower has a built-in ventilator which separates the pellicles from the kernels directly. Throughput: 90kg per hour. Dimensions: 125 × 75 × 48 cm.

PRICE CODE: 2

UNATA 5202 PEANUT DECORTICATOR

This decorticates and winnows in one process. Peanut pods are brought in via a funnel and broken by the pressure generated between a rotating drum and a counter drum composed of simple metal bars. When the hulls are broken, a fan separates them from the nuts. Throughput: 60–120kg per hour. Manual and motor-driven models are available.

PRICE CODE: 2

UNION FOR APPROPRIATE TECHNICAL ASSISTANCE
G. Van den Heuvelstraat 131
3140
RAMSEL
BELGIUM

Manual decorticators

Commodity	Country	Manufacturer	Equipment name	Price code	Throughput (kg per hour)	Other information
Maize	Belgium	Ateliers Albert & Co. SA	CORN-SHELLER R250	2	250–300	Optional electric power
		Union for Appropriate Technical Assistance	UNATA 5200	2	90	
	India	Dandekar Brothers – Engineers	HAND-OPERATED MAIZE SHELLER	1		
		Cossul & Co. Pvt Ltd	HAND MAIZE SHELLER	1	37–50	
			HAND & POWER-OPERATED MAIZE SHELLER	1	100–120	Throughput with power is 200–300kg per hour
		Rajan Universal Exports (MFRS) Ltd	MAIZE SHELLER MASTER	1	35–2500	Other models are available
	Ghana	Agricultural Engineers Ltd	HAND-OPERATED MAIZE SHELLER	1	250	
	Japan	Cecoco	CORN SHELLER	2	100–150	
			CORN SHELLER	2	300–350	
	Liberia	Agro Machinery Ltd	HAND-OPERATED MAIZE SHELLER	1	250	
			HANDY MAIZE SHELLER	1	50–120	
	Malawi	Ministry of Agriculture	CHITEDZE HAND-OPERATED MAIZE SHELLER	1	30	
	Nigeria	Institute for Agricultural Research	MAIZE SHELLER, MANUAL	1	13–49	Commercially unavailable
	Philippines	Agricultural Mechanization Development Programme	ITA CORN SHELLER	1	22–30	
	Tanzania	Tanzania Engineering & Manufacturing Design Organization	MAIZE SHELLER	2	500	Has capacity for tractor power
	UK	Alvan Blanch Development Co. Ltd	MAIZE SHELLER	1	50	
	USA	Almaco	MANUAL EAR CORN SHELLER	1	100	
		Lehman Hardware & Appliances Inc.	CORN OFF	1		
	Zimbabwe	G. North & Son (Pvt) Ltd	HAND-DRIVEN MEALIE SHELLER	1		Sheller weighs 6kg
		Multi Spray Systems	HAND-OPERATED MAIZE SHELLER	1		Sheller weighs 200g
Groundnuts (Peanuts)	India	Dandekar Brothers – Engineers	BABY STYLE GROUNDNUT DECORTICATOR	1	50	Can be electrically operated using 0.75kW motor
	Belgium	Union for Appropriate Technical Assistance	UNATA 5202 PEANUT DECORTICATOR	2	60–120	Other manual and motor-driven models are available
	Philippines	Agricultural Mechanization Development Programme	AMDP HAND-OPERATED PEANUT SHELLER	1	18–20	Can be converted to pedal drive
Cashew nuts	Thailand	Department of Agriculture	CASHEW NUT SHELLER	1		
Walnuts	USA	RR Mill Inc.	BELLCORN SHELLER	1		Also shells dry ear corn
		Lehman Hardware & Appliances Inc.	NUT CRACKER	1		Also cracks other small nuts
Peas	USA	Lehman Hardware & Appliances Inc.	TEXAS PEA SHELLER	1		Shells dream, blackeye, and purple hull peas. Can also be motorized

(continued)

Decorticators

Manual decorticators (continued)

Commodity	Country	Manufacturer	Equipment name	Price code	Throughput (kg per hour)	Other information
Longan fruits	Thailand	Chiangmai Commerce Organization	LONGAN HULLER	1		
Cocoa	UK	John Gordon & Co. Engineers Ltd	COCO BREAKING/ WINNOWING SYSTEM	2		Machine can be supplied with electric motors
Coffee	UK	John Gordon & Co. Engineers Ltd	NO. 10 AFRICA HULLER	2	36	Can also shell dry cherry (22kg/h)
			BUKOBA COFFEE HULLER	2	9	Shells dry and parchment
Rice	Japan	Cecoco	TWO–MAN HAND HULLER	2	250	Also hulls sunflower seeds (10–150kg/h)
			HAND RICE POLISHER	1	10–15	Power drive available
	UK	John Gordon & Co. Engineers Ltd	JAVA RICE HULLER	1	14	
	UK	Alvan Blanch Development Co. Ltd	HAND HULLER	1	15	

Powered decorticators

Commodity	Country	Manufacturer	Equipment name	Through-put kg per hour	Price code	Power (kW)
Maize	Brazil	Messrs. Laredo	CORN HUSKER/SHELLER	60	2	1.5
	Colombia	Penagos Hermanos & CIA. Ltda	MAIZE SHELLER	300–600	3	0.3
	Ghana	Agricultural Engineers Ltd	HANDY MAIZE SHELLER	50–120	2	
	India	Mohinder & Co. Allied Industries	MAIZE SHELLER	1000–1500	3	2.2
		Rajan Universal Exports (MFRS) Ltd	MAIZE SHELLER	400–600	3	1.5
	Japan	Cecoco	CORN SHELLER	250–300	3	3.7
	Nigeria	Institute for Agricultural Research	MAIZE DEHUSKER/SHELLER	300–400	2	3
Grains (various)	Belgium	Union for Appropriate Technical Assistance	DECORTICATOR	50–150	2	
	Zimbabwe	ENDA Zimbabwe (Pvt) Ltd	MINI DEHULLERS		4	
Groundnuts (Peanuts)	India	Dandekar Brothers (Engineers & Co)	'BABY STYLE' GROUNDNUT DECORTICATING MACHINE	50	2	0.7
	Japan	Cecoco	PEANUT HUSKER/SHELLER	40–50	2	
	Nigeria	Institute for Agricultural Research	GROUNDNUT AND COWPEA SHELLER	250	3	2.5
Palm nuts	Ghana	Agricultural Engineers Ltd	PALM NUT CRACKER	300–500	2	3.7
	India	Rajan Universal Exports (MFRS) P Ltd	MOBILE PALM OIL NUT CRACKER	500	2	1.1
Pulses (various)	India	Dandekar Machine Works Ltd	DAL SCOURER	750–950	3	3.7
		Gardeners Corporation	PEA HULLERS		3	0.7
	Japan	Cecoco	BEAN DEHUSKING & SPLITTING MACHINE	180–240	3	3.7
Sunflower seeds	UK	Lewis C. Grant Ltd	SUNFLOWER HULLER No. 15D	1250	4	4.1

(continued)

Powered decorticators (continued)

Commodity	Country	Manufacturer	Equipment name	Through-put kg per hour	Price code	Power (kW)
Sorghum/ millet	Botswana	Rural Industries Innovation Centre	TSHILO/DEHULLER (SORGHUM DEHULLER)		1	
Rice	India	Dandekar Machine Works Ltd	RUBBER ROLL SEALER	900–1100	3	2.2
		Devraj & Company	RUBBER ROLL SHELLER		2	2–5
			EMERY CONE POLISHER	300	3	3.7
			DEVRAJ DISC POLISHER	350–400	3	7.5
			JM5 (V BELT) TYPE	500–700	2	3.7
			JM5 (KING) GEAR TYPE	800–1000	3	3.7
		Kisan Krishi Yantra Udyog	CENTRIFUGAL PADDY DEHUSKER	400–500	3	1.5
			RUBBER ROLL SHELLER WITH HUSK ASPIRATOR	500–700	3	3.7
		Northern India Flour Millers Corporation	MODERN RICE HULLER	200–300	3	3.7
	Japan	Cecoco	HORIZONTAL ABRASIVE ROLL TYPE RICE POLISHER	300–1200	4	3.7
			AUTO RUBBER ROLL RICE HULLER	360	4	0.4
	Thailand	Ruang Thong Machinery Lp.	RUANG THONG RICE HULLER	50	2	1.5–2.2
				100	2	3–3.7
				200	3	6
			PIN THONG RICE MILL	200–250	3	9
	UK	Lewis C. Grant Ltd	HULLER WITH FABRICATED STAND 2C	230–260	3	11
			2C HULLER COMBINE	230–260	3	11
			NO. 2 MAIZE HULLER	180	3	11–15
			1 RICE HULLER & POLISHER	230–290	4	11
		Mackies Pty Ltd	BON ACCORD RICE HULLERS	80–130	3	3.7–4.5
				300–400	3	6.7–8.9
Coffee	Brazil	INCOBI	RICE AND COFFEE PROCESSOR		2	3
	India	Kemajuan	HULLER E303	100–250	2	3.7
		Rajan Universal Exports (MFRS) Ltd	COFFEE HULLER	125–185	2	3.7
	UK	Denlab International (UK) Ltd	COFFEE HULLER AH4	210	3	3.7
		John Gordon & Co. (Engineers) Ltd	AFRICA HULLERS	700	4	15
		Lewis C. Grant Ltd	NO. 5 AFRICA HULLER	114–173	4	2.2–3
		Mackies Pty Ltd	MACS AFRICA HULLERS	170	3	2.2–3
				325	3	4.5–6

'Baby style' groundnut decorticator

UNATA 5202 peanut decorticator

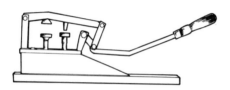

Lehman nutcracker

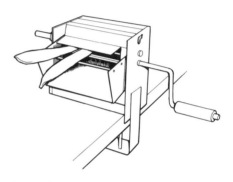

'Texas' pea sheller

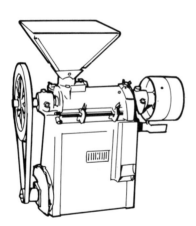

2C Huller combine

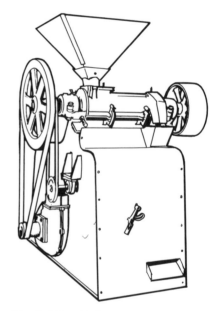

No. 5 Africa Huller

20.0 Depositors

These pieces of equipment are used for the moulding and forming of specific products such as bread, biscuits, pies and confectionery.

'TRIPLEX' COOKIE FORMER

This deposit and high-speed wire-cutting machine enables a wide range of biscuits to be produced. All operating controls are located on one side of the machine. Machine speed, feed run speed, length of wire stroke and band height are adjustable while the machine is running. A clutch is provided to disconnect the wire cutting mechanism while running continuous bar goods. The wire-cutting mechanism can make up to 250 cuts per minute.

PRICE CODE: 4

BAKER PERKINS
Westwood Works
Westfield Road
PETERBOROUGH
UK

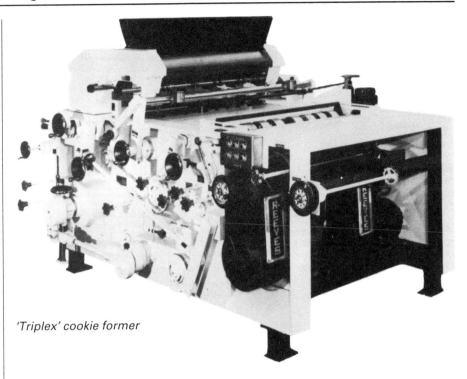

'Triplex' cookie former

21.0 De-stoners

CHERRY PITTER

This plastic, hand-operated pitter removes the stones from cherries. A punch action forces the stone through the cherry and into a waste container. The cherry is left with a small hole in the centre and drops into a separate bowl. The unit clamps to any surface between 1.6 and 4.1cm thick. Throughput is 2520 cherries per hour. Height: 19cm.

Other models include the Traditional Cherry Pitter and the Faster Cherry Pitter.

PRICE CODE: 1

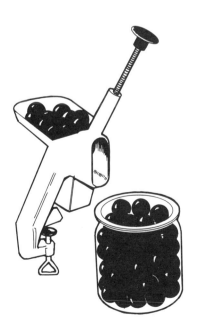

LEHMAN HARDWARE & APPLIANCES INC.
P.O. Box 41
4779 Kidron Road
Kidron
OHIO 44636
USA

Similar models, but motor-driven rather than manual, are also available from Lehman Hardware.

22.0 Distillation equipment

Manufacturers of glassware for distillation:

QVF/CORNING PROCESS SYSTEMS
Corning Ltd
STONE
Staffordshire
ST15 OBG
UK

23.0 Dryers

23.1 Solar dryers

Drying foods in the sun is probably the oldest method of food preservation. At the simplest level, the food is spread on the ground or on mats, and the moisture at or near the surface of the food is heated and vaporized by heat from the sun and ambient air. However, there are a number of disadvantages with sun drying. These include: the intermittent nature of solar energy throughout the day and different times of the year; the possible contamination of the food material by dirt, or insects; exposure to the elements (such as rain and wind), causing spoilage and losses; and the exposure of the crops to rats, chickens and other pests.

Most solar dryers incorporate a platform, raised above the ground, and often some kind of covering to combat the problems mentioned above. The purchase of a solar dryer may require a high capital investment initially, but it is argued that the gains from producing a higher-quality product will quickly offset these costs. A careful assessment of the requirements of the producer is therefore necessary to establish whether higher income can be obtained from better quality products. Solar dryers may be more suitable if processing high value foods rather than low-value staple crops. In addition, the introduction of combination solar/fuel fired dryers may be a desirable option to overcome the problems of intermittent sunshine that often adversely affects solar drying.

SOLAR FOOD AND CROP DRYER

This dryer is not commercially available. Two wooden boxes are made, one to fit inside the other. The inner box should have a 6cm gap all the way around for holding insulation. A few trays, with mesh bases are required to hold the material for drying, and ventilation holes are required all around the sides of the box. Plastic sheeting is used to form a clear lid over the dryer which retains heat inside the unit. Dimensions: 2.6 × 1.2 × 0.4m. Information available from:

ILO - SDSR
P.O. Box 60598
NAIROBI
KENYA

Solar dryers

FORCED CONVECTION SOLAR CROP DRYER

This dryer is not commercially available. It has a centrifugal/axial flow fan to blow air through the drying chambers. Gas, electric and firewood heaters are also available for combination dryers. Information from:

INSTITUTE FOR AGRICULTURAL RESEARCH
Samaru, Ahmadu Bello Uni
PmB 1044
ZARIA
NIGERIA

TRIPLE PASS SOLAR COLLECTOR/ DRYER

This dryer can be used for cereals and tubers and consists of a flat plate collector, drying cabinet and dehumidification chamber. These units result in high heat-gain through the collector, and sustained and uniform drying for both day and night-time.

PRICE CODE: 2

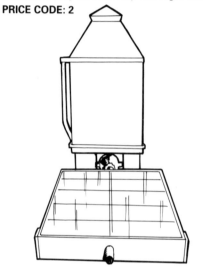

DEHUMIDIFIED AIR SOLAR-HEATED TRAY DRYER

This dryer comprises several built-in modules each consisting of a flat plate collector, dehumidification chamber and drying cabinet. All major components are detachable and so can be added to, increasing the overall capacity. A wind-operated suction pump draws air through, and temperatures of up to 90°C can be reached. Throughput: 200kg (fish), in four days.

PRICE CODE: 3

DEPARTMENT OF AGRICULTURAL ENGINEERING
Faculty of Engineering
University of Nigeria
NSUKKA
NIGERIA

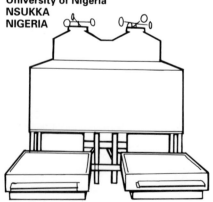

Solar dryers also manufactured by:
SCIENTIFIC INFORMATION DIVISION
Pakistan Research Council of
Scientific & Industrial Research
39 Garden Rd
KARACHI 0310
PAKISTAN

M/S DALTRADE (NIG) Ltd
Plot 45 Chalawa Indust. Est.
P.O. Box 377
KANO
NIGERIA

23.2 Fuel-fired dryers

CARDAMOM DRYER

This machine comprises a drying bin, and a heating chamber powered by an electric fan which requires motors, 0.55kW (0.74hp) 1500rpm single-phase. The heaters require 8kW. Capacity: 1000kg per charge. Dimensions: 930 × 350 × 250cm.

The cardamon dryer is not commercially available. Information from:

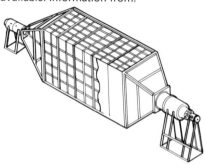

TANZANIA ENGINEERING & MANUFACTURING DESIGN ORGANIZATION
P.O. Box 6111
ARUSHA
TANZANIA

KILN DRYER

This kiln dryer comprises a drying bin, furnace, and air circulatory systems, and will run continuously for 24 hours. It is used to dry onions at a throughput of 3.3 tonnes/3 days. Dimensions: 360 × 327 × 488cm.

PRICE CODE: 4
FOREST PRODUCTS & RESEARCH DEVELOPMENT INSTITUTE
Officer in Charge, Packaging Prg
LAGUNA
PHILIPPINES

DRYER

A general-purpose dryer (for paddy and other grains) can be manufactured locally. The power source can be a petrol/kerosene engine, electric motor or a rice-husk burner. This dryer is not commercially available. Capacity: 1 tonne.

INTERNATIONAL RICE RESEARCH INSTITUTE
PHILIPPINES

BATCH DRYER

A rice dryer, based on an IRRI design, uses a 3hp (2.2kW) petrol engine or a 1.5kW (2hp) motor to drive a 47cm diameter vane-axial type fan and to supplement heating. Primary heat is provided by a rice-hull fur-

nace or kerosene burner. There is an automatic safety burner shut-off mechanism. The weight (without engine) of the burner and fan is 40kg. Airflow is 1800 cfm at 2200 rpm, with a static pressure of 20mm water. Temperature: 43°C. Fuel consumption: 0.75 litres per hour petrol + 2.0 litres per hour kerosene or 3.4kg/h rice-hull. Drying bin dimensions: Length 277cm, width 190cm, height 92cm. Drying rate 23% moisture (w.b.) paddy (1 tonne) to 14% in 5–6 hours.

PRICE CODE: 3

VERTICAL BIN BATCH DRYER

This rice dryer is constructed of wood and steel and uses a 2.2kW (3hp) motor or a 3.7kW (5hp) engine. Heat is supplied by a kerosene burner. Consumption: 1.5 litres per hour petrol + 2.7 litres per hour kerosene. Air at 43°C is blown by the blower (2000 rpm) through a grain bed 46cm thick at a flow rate of 1612 cubic metres per hour (3600 cfm) and 3cm water static pressure. Weight with engine: 364kg. Dimensions: length 344cm, width 173cm, height 158cm. Capacity: 2 tonnes paddy per load. Drying rate: 2% per hour.

PRICE CODE: 3
JCCE INDUSTRIES
242 Mayondon
Los Banos
LAGUNA
PHILIPPINES

MULTI-CROP DRYER, TILTING BED TYPE

This tilting bed dryer is manually-operated and uses rice husk as a fuel. The 1500 rpm multi-vane centrifugal fan gives an airflow of 3500cfm. In addition, a fixed-flat bed type is available (Model FS 35).

PRICE CODE: 2

MULTI-CROP DRYER, FIXED TWIN-BED TYPE

This dryer is indirectly heated from rice husk. It comprises a burner, blower and two holding beds totalling 90–100 cavans. Each bed can be operated simultaneously or alternately. Model FD50 has a capacity of 45–50 cavans/bed × 2. Throughput: 15–17 cavans per hour. Size range: 30–35 cavans/bed × 2. (Model FD35)

PRICE CODE: 0
MARINAS MACHINERY MANUFACTURING CO. INC.
Rizal Street
Pila
LAGUNA
PHILIPPINES

TRAY DRYER

This dryer has proved successful in drying many products, particularly the higher value foods such as fruit and herbs.

It consists of sixteen trays, each measuring 3ft by 4ft. The capacity of the dryer varies according to the product being dried, but as an example, it will dry 300–400kg of herbs (net weight) per day.

The heat source can be powered by diesel, gas or electric. The dryer requires 1200 cubic feet of hot air per minute.

Work is continuing on the dryer in an effort to reduce the cost of the forced hot air system.

Plans, construction guide and a video (in Spanish) are all available from ITDG.

ITDG
Myson House
Railway Terrace
RUGBY
Warwickshire CV21 3HT
UK

Fuel fired dryers also manufactured by:
ARMFIELD TECHNICAL EDUCATION CO. LTD
Bridge House
West Street
RINGWOOD BH24 1DY
UK

CECOCO
P.O. Box 8
Ibaraki City
Osaka 567
JAPAN

TECHNOLOGY DEVELOPMENT OFFICE
S & T Comm'n of Liaoning Province
1–2 San Hao Street
He Ping District
SHENYANG
CHINA

AGRICULTURAL MECHANIZATION DEVELOPMENT PROGRAMME
AMDP, CEAT, UP Los Banos College
LAGUNA, 4031
PHILIPPINES

C.V. KARYA HIDUP SENTOSA
Jl. Magelang 144
YOGYAKARTA, 55241
INDONESIA

P.T.RUTAN MACHINERY TRADING CO.
P.O. Box 319
SURABAYA
INDONESIA

23.3 Electric dryers

SUN PANTRY OVEN DRYER

This drier is designed for small-scale drying, and uses heat from an oven (temperature range of 41°–71°C). The individual trays can also be used for sun-drying. The dryer is made from hardwood and comes with four drying trays which have aluminium screens, covered in acrylic for safe contact with the food. Drying space: 4955cm². Dimensions: height 17 × weight 43 × depth 33cm.

PRICE CODE: 1

LEHMAN HARDWARE & APPLIANCES INC.
P.O. Box 41
4779 Kidron Road
Kidron, OHIO 44636
USA
DRYER

This electric or gas dryer, can be made to order. It is made of stainless steel and has double doors at the front.

PRICE CODE: 2–3

KLAUY NAM THAI THOW OP
1505 07 Rama 4 Rd, Wangmai
Patumwan
BANGKOK 10330
THAILAND

Electric dryers also manufactured by:
LEHMAN HARDWARE & APPLIANCES INC.
P.O. Box 41
4779 Kidron Road
Kidron, OHIO 44636
USA

23.4 Vacuum dryers

ROTARY VACUUM DRYERS

This vacuum dryer is suitable for producing granular, powdery materials. It has a capacity of 250 litres and gives low temperature, uniform drying.

PRICE CODE: 4

DALAL ENGINEERING PVT LTD
36–37 Jolly Maker Chambers II
Nariman Point
BOMBAY 400 021
INDIA

23.5 Spray dryers

SPRAY DRYER

This dryer produces powdered foods and is designed for product testing or small-scale production of high-value foods. The con-centrate feed-rate may be varied using a peristaltic pump, and the temperature of the drying air is controlled by a 3kW heater.

PRICE CODE: 4

ARMFIELD TECHNICAL EDUCATION CO LTD
Bridge House
West Street
RINGWOOD BH24 1DY
UK 01425 478781

24.0 Enrobers

These continuous machines are used to coat foods in chocolate, batter or other coating materials. The food passes on a wire conveyor belt through a curtain of coating material.

PRE-BOTTOMING ENROBER

This pre-bottoming enrober will handle all types of chocolate and other coating substances and is complete with an automatic tempering system. The main body of the machine is made from stainless steel. It comes complete with water cooled table/tables and band drive. All electrical controls are built into the machine. During operation untempered chocolate is held in the lower tank and is pumped by the vertical screw through the tempering tube to the stirred bottomed bath.

PRICE CODE: 4

BAKER PERKINS
Westwood Works
Westfield Road
PETERBOROUGH
UK

Enrobers also manufactured by:

SOLLICH GMBH & CO. KG
D -4902
BAD SALZUFLEN
GERMANY

ROBINSONS OF DERBY
DERBY
UK

Pre-bottoming enrober

25.0 Evaporators

Plate evaporators are used in the process of evaporation. They are designed to minimize residence time at processing temperatures and thus minimize damage to heat-sensitive products.

Plate evaporators are expensive and are not recommended for most small-scale processing operations.

Plate evaporators are available from:

APV EVAPORATORS
APV Company Ltd
CRAWLEY
West Sussex
UK

26.0 Expellers

These machines, which are similar to extruders, are used for extracting oil from oilseeds or nuts. They consist of a horizontal cylinder which contains a rotating screw. The seeds or nuts are carried along by the screw and the pressure is gradually increased. The material is heated by friction and/or electric heaters. The oil escapes from the cylinder through small holes or slots and the press cake emerges from the end of the cylinder. Both the pressure and temperature can be adjusted for different kinds of raw material.

TINY OIL MILL

This is a variable pitch screw expeller, capable of up to 73% extraction rates on a single pass of material. Up to 100kg per hour of groundnut and 80kg per hour of sunflower seed can be processed. The expeller is fitted with a steam kettle for preheating the seed. Optional extras include a boiler for the steam kettle which uses the husk as a fuel, filter press, filter pump and ground nut decorticator. The complete plant is driven via a belt and pulleys from a line shaft powered by a 7.5kW three phase electric motor or a 10kW diesel engine.

PRICE CODE: 2

TINYTECH PLANTS PRIVATE LTD
Near Mallavia Wadi
Gandal Road
RAJKOT - 360 002
INDIA

27.0 Extruding machines

27.1 Hot extruders

EXTRUDER TYPE IIB265

This machine makes extruded products such as snack foods from cereals. The food used to make the extruded product is fed into the hopper and milled where it is mixed to form a dough paste mixture. The substance is then fed through the heated extrusion unit and is shaped into the required form. Dimensions: 1480 × 2570 × 1100mm. Smaller models are available: EXTRUDER R32–S35, and EXTRUDER II B275.

PRICE CODE: 4

PABRIK MESIN
P.T. Mata Hari SS
JL Cempala 18
INDONESIA

27.2 Cold extruders

SAUSAGE MAKER KIT

This hand-operated sausage kit is designed for use on a small scale. The mixture is placed inside the gun and extruded into sausage skins through a nozzle. Included in the kit are standard and chippolata nozzles, premix, sausage skins and a recipe book.

PRICE CODE: 1

LAKELAND PLASTICS
38 Alexandra Buildings
WINDERMERE
LA23 1BQ
UK

ROTARY TYPE EXTRUDER

This hand-operated stainless steel extruder consists of two main components: a hollow body fitted with an interchangeable die plate, and a lid fixed to a movable central screw.

PRICE CODE: 1

HAND-HELD EXTRUDER

This is a hand-operated piston-type extruder for dough-like materials, and is available in many variations in Asia. It can be fabricated from metal or wood, with either a fixed or interchangeable die plate. Dimensions: 8 × 25 × 7cm.

PRICE CODE: 1

ANJALI KITCHEN COMPONENTS
Mahavir Brothers
202 Trivani House
Kelivadi Back Road
BOMBAY
400004
INDIA

Powered oil expellers

Commodity	Country	Manufacturer	Equipment name	Through-put kg per hour	Price code	Power (kW)
Coconut	Philippines	Sev Corporation	OIL EXPELLER	50	4	
Groundnut	Ghana	Agricultural Engineers Ltd	AGRICO OIL EXPELLER: CAPTAIN MODEL	512	3	
Sunflower	Tanzania	Tanzania Engineering & Manufacturing Design Organization	OIL EXPELLER EXP 10	110	4	4
Oil seeds (various)	Germany	IBG Monforts & Reiners Gmbh & Co.	KOMET CA59	8–15	4	0.37
			KOMET DD85	25–70	4	1.75/3.5
			OIL EXPELLERS	2–5	4	
	Japan	Cecoco	OIL EXPELLER (3 models)	30–50	3	
				50–80	3	
				80–100	4	
	Tanzania	Tanzania Engineering & Manufacturing Design Organization	OIL EXPELLER EXP 20	110	3	7.5

28.0 Filling machines

The four basic methods for liquid filling are: vacuum filling, measured dosing, gravity filling and pressure filling. Filling by vacuum is the cleanest and most economical way to fill many products ranging from low to high viscosities, whereas gravity and pressure filling are best suited to the moderately rapid filling of low viscosity liquids such as fruit juices, which flow easily.

28.1 Liquid fillers

Manual liquid fillers

LIQUID FILLER

This filler fills different sized bottles with liquids and is operated by two people. There are 14 nozzles of 13 or 15mm diameter, each with a flow rate of 57 litres per hour. Dimensions: $75 \times 74 \times 100$cm.

PRICE CODE: 3

YUNAN GENERAL MACHINE WORKS
24 Dadi Lu
Duncheng Zheng
Yunan County
Guangdong Province
CHINA

10500 DANA API MANA

This hand-operated filling unit for honey consists of a free-standing frame fitted with a plastic tank. The honey is gravity fed into the containers. Throughput: 300–400 jars per hour.

PRICE CODE: 2

SWIENTY
Hortoftvej 16
Ragebol
Sonderborg DK 6400
DENMARK

HAND-OPERATED DISPENSER

This equipment can dispense two products into one container. In addition, different volumes can be dispensed, as the pumps are adjustable.

PRICE CODE: 3

ADELPHI MANUFACTURING CO. LTD
Olympus House
Mill Green Road
HAYWARDS HEATH
RH16 1XQ
UK

JAM FUNNEL

This stainless steel funnel is used for filling jars, and has a splash guard around the top.

PRICE CODE: 1

LAKELAND PLASTICS
38 Alexandra buildings
WINDERMERE
LA23 1BQ
UK

ASPIRATOR, HEAVY DUTY POLYTHENE

These liquid containers each have a tap for filling. They also have clamping handles and a screw cap closure for the necks. Size range: 2.5–50 litres.

PRICE CODE: 1

THE NORTHERN MEDIA SUPPLY LTD
Sainsbury Way
HESSLE
HU13 9NX
UK

GRAVITY FILLER

This machine is used for filling liquids, operating on a vertical plunger principle. Hoppers are available in several sizes to meet varying production requirements. The operator holds the container under the nozzle and the filler handle dispenses up to 454g per stroke. The quantity can be adjusted by a clamp block on the piston rod. Dimensions: $61 \times 61 \times 102$cm.

PRICE CODE: 2

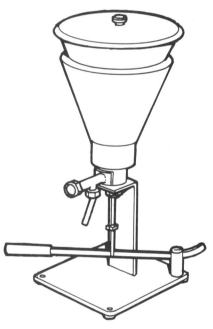

FMC CORPORATION
Machinery International Division
P.O. Box 1178
SAN JOSE
California 95108
USA

STRAIDLE

This ladle is made from heavy duty stainless steel and is equipped with a convenient strainer that swings under the pouring liquid. Alternatively, the strainer can be moved for 'strainless' pouring. Length: 35.6cm.

PRICE CODE: 1

LEHMAN HARDWARE & APPLIANCES INC.
P.O. Box 41
4779 Kidron Road
Kidron
OHIO 44636
USA

Funnels also manufactured by:

LEHMAN HARDWARE & APPLIANCES INC.
P.O. Box 41
4779 Kidron Road
Kidron
OHIO 44636
USA

Fillers also manufactured by:

ADELPHI MANUFACTURING CO. LTD
Olympus House
Mill Green Road
HAYWARDS HEATH
RH16 1XQ
UK

Powered liquid fillers

VOLUMETRIC FILLING MACHINE

This is a synchronized filling machine for cans, cartons etc. There are two filling outlets with a filling volume per stroke of 0–5kg.

PRICE CODE: 2

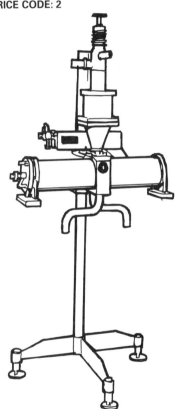

GERSTENBERG OG AGGER A/S
Frydendalsvej 19
1809 FREDERSIKSBERG
DENMARK

SEMI-AUTOMATIC FILLER D1000

This is used for filling liquid or semi-liquid products. It is operated by a foot-activated switch, and the filling volume for one stroke can be adjusted. Max quantity per stroke = 1 litre.

The models available are: DA 1000 & DOUBLE D1000.

PRICE CODE: 3

PRIMODAN FOOD MACHINERY A/S
Oesterled 20–26
P.O. Box 177
DK-4300 HOLBAEK
DENMARK

Filling machines

ADELPHI SEMI-AUTOMATIC LIQUID FILLER

This filler is used for the non-measured filling of a variety of liquids (e.g. spirits, oils and syrups) into most types of bottles. It can either be gravity fed from a container, or a mixer can be fitted with a feed pump. The stainless steel filling valve is fitted with a 0.95cm diameter nozzle. Other sizes are available.

PRICE CODE: 2

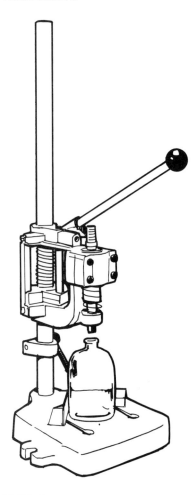

ADELPHI MANUFACTURING CO. LTD
Olympus House
Mill Green Road
HAYWARDS HEATH
RH16 1XQ
UK

VACUFIL

This automatic vacuum bottle filler is fitted with two heads so that two bottles can be filled at the same time. There is also a glass reservoir flask to collect excess liquid. The pump is fitted to a 0.19kW (0.25hp) motor, suitable for 220V AC. Throughput: 600 bottles per hour.

PRICE CODE: 3

J T JAGTIANI
National House
Tulloch Road
Apollo Bunder
BOMBAY 400 039
INDIA

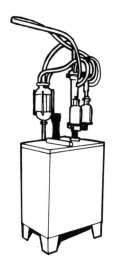

MASTER ROTARY MULTIHEADS GRAVI-FILL

This machine fills containers with any free-flowing liquid. It has a stainless steel overhead rotating tank fitted with a float valve inlet, which is connected to a main supply tank. The machine has a variable speed drive mechanism (power: 0.75kW/1hp) and a 'no bottle, no fill' mechanism. Dimensions: length 244 × width 122 × height 244cm.

PRICE CODE: 3

MASTER MECHANICAL WORKS PVT LTD
Pushpanjali SV Road
Santa Cruz (W)
BOMBAY 400 054
INDIA

LIQUID & VISCOUS MATERIAL FILLERS

Model PSF1 is a pneumatically-operated filling machine, comprising a hopper with a 40 litre capacity. The machine itself is operated by a foot switch for a single or contin-

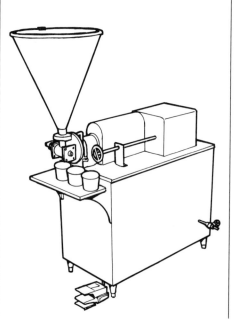

uous operation and has a filling range of: 50–100ml or grams, 100–250ml or grams, 250–1000ml or grams. Throughput: 10–15 containers/min.

PRICE CODE: 3

PANCHAL WORKSHOP
P B No 62, Anand Sojitra Road
ANAND 388 001
INDIA

SMALL-SCALE BOTTLING EQUIPMENT

This equipment comprises a mixer/dispenser, a crown corker and a pasteurizer and can be used for most liquids including palm sap and high acid juices (pH 1.5–4.6). Throughput: 200 × 0.29l crown corked bottles per hour.

PRICE CODE: 4

NIGERIAN INSTITUTE FOR OIL PALM RESEARCH
Private Mail Bag 1030
BENIN CITY
NIGERIA

PISTON FILLER, MODEL A

This machine will fill a wide range of jars or plastic containers with liquids or semi-liquids (such as ice cream, honey and cottage cheese) in predetermined amounts from 28.3 grammes to 936 grammes. It comprises a 45.5-litre cone hopper and a positive piston filler driven by a 0.19kW (0.25hp) motor. The containers are fed under a discharge spout to achieve a bottom fill. Dimensions: 58 × 102 × 86cm. Versions available include: Table, Twin, Gallon models.

PRICE CODE: 3

FMC CORPORATION
Machinery International Division
P.O. Box 1178
SAN JOSE
California 95108
USA

Liquid fillers also manufactured by:

DAIRY UDYOG
C-229A/230A, Ghatkopar Industrial Estate
L.B.S. Marg, Ghatkopar (W)
BOMBAY 400 086
INDIA

PANCHAL WORKSHOP
PB No 62
Anand Sojitra Road
ANAND 388 001
INDIA

28.2 Solid fillers

Two basic methods are used for filling powders, granules and high viscosity products. They are volumetric filling and filling after weighing. The nature of the product usually determines which method is used. Volumetric filling includes filling by auger, flask fillers and vacuum filling. Filling by weight is perhaps the most satisfactory way of meeting the requirements of Weights and Measures Acts or similar legislation, as the product is weighed before being filled.

Manual solid fillers

HAND-OPERATED JAM FILLER

This filler is suitable for filling jars and cans. It comprises a hopper of 20kg capacity and can be fitted with either a vertical or horizontal filling attachment, both capable of filling from 50–500 grammes per lever stroke.

PRICE CODE: 2

RAYLONS METAL WORKS
Kondivitta Lane
Andheri-Kurla Road
BOMBAY 400 059
INDIA

Powered solid fillers

DOSING UNIT

This machine is used for dosing viscous products.

The product is fed into a funnel and via the transfer valve and dosing nozzle into the container. The stainless steel unit has an adjustable filling temperature and a dosing volume of 80–500ml and a filling temperature of 5–60°C (optional up to 90°C). The doser may be stopped at any time by a lever which controls the transfer valve. A support table is available as an optional extra. Power requirements: 220/380V.

PRICE CODE: 3

BRAS HOLANDA
P.O. BOX 1250
CURITIBA PR-80.001
BRAZIL

AUTO HYDRAULIC FILLER/STUFFER

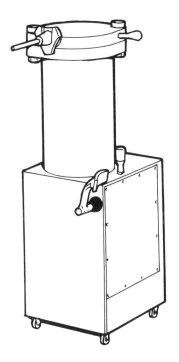

This machine is used for producing sausages, and is available with single-phase or three-phase motors. It operates when the knee lever is pressed, leaving both hands free. The piston has two gaskets to prevent the product from escaping between the

joints. Three nozzle sizes are available: 13, 20, 30mm diameter. Capacity: 15 litres or 12kg.

PRICE CODE: 2

TALSABELL S.A.
Poligono Industrial V
Salud 8-46.950
VALENCIA
SPAIN

WEIGH FILLER, MODEL GBF 20

This semi-automatic bulk filling machine operates by gravity feed. Two containers are placed on pans beside each other and are then filled alternately until they weigh the required amount. The machine has a filling range of: 5, 10, 15 and 20kg, with a throughput of 3–5 fills per minute.

PRICE CODE: 2

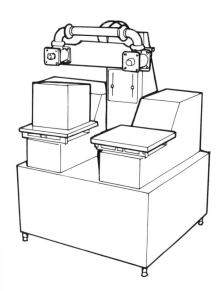

PANCHAL WORKSHOP
P B No 62, Anand Sojitra Road
ANAND 388 001
INDIA

FILLING MACHINE

This semi-automatic filling machine is suitable for pulp, jam and jelly. It has a power requirement of 0.56kW (0.75hp) 110/220v AC with a hopper capacity of 20–30 litres.

Throughput: 1500 fillings per hour. Dimensions: $1.60 \times 0.53 \times 0.53$m.

PRICE CODE: 4

VILLAMEX S.A DE C.V
Sur 69-A No. 407
Col Bandjidal
MEXICO DF 09450
MEXICO

Solid fillers also manufactured by:
FRAGOL LTDA
Av Libertador Brig Gral
Lavalleja 1641-Esc 203
MONTEVIDEO
URUGUAY

RAYLONS METAL WORKS
Kondivitta Lane
Andheri-Kurla Road
BOMBAY 400 059
INDIA

28.3 Paste fillers

MULTI-PURPOSE PASTE FILLING MACHINE

This is used for filling tubes, bottles, cans etc. and is suitable for measures between 5g and 100g. All contact parts, including the hopper, are made from stainless steel.

PRICE CODE: 2

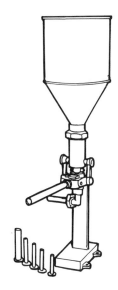

J T JAGTIANI
National House
Tulloch Road
Apollo Bunder
BOMBAY 400 039
INDIA

SAUSAGE FILLERS

This unit is made from stainless steel and is available in two sizes (5 litres and 10 litres capacity). Different-sized nozzles are also available.

PRICE CODE: 2

In addition, a *hydraulic filler* for sausages and pastes is available from:

STADLER CORPORATION
P.O. Box 7177
BOMBAY
400 070
INDIA

28.4 Powder fillers

FORM FILL SEAL MACHINE

This machine packages powders, granules and particulate products. Optional features include: batch coder, perforation and a notching unit.

PRICE CODE: 3

KAPS ENGINEERS
831 GIDC
Makapura
VADODARA 390 010
INDIA

ALITE LF 6, POWDER FILLING MACHINE

This single-head powder filling machine has a stainless steel loading hopper. It is controlled by a foot pedal. The vee belt drive is from a 0.75kW (1hp) motor and is mounted on a stand.

PRICE CODE: 3

WINKWORTH MACHINERY LTD
Willow Tree works
Swallowfield Street
Swallowfield
READING RG7 1QX
UK

POWDER FILLING SYSTEMS

This is suitable for small volume filling. A screw displaces a measured quantity of powder and revolutions of the auger are sensed by photo-electric pulses. The thumbwheel selectors can change the fill quantity without stopping the machine. Throughput: 900 to 2400 fills per hour.

PRICE CODE: 4

SQUARE TEC
35 Industrial Estate
LONAVLA 410 401
INDIA

AUTO FILLER FOR POWDERS

This machine fills bottles and cans with powders. It has an infinitely variable speed drive. Throughput: 240–600 containers per hour per head. Hopper capacity: 50 litres. Power requirements: 2.2kW.

PRICE CODE: 4

RECON MACHINE TOOLS PVT LTD
37 Sarvodaya Industrial Estate
Mahakali Caves Rd, Anderi (E)
BOMBAY 400 093
INDIA

FILLING MACHINE

This stainless steel machine is designed for powders.

PRICE CODE: 1

INDUSTRIAS TECHNICAS 'DOLORIER'
Alfredo Mendiola 690
urbanizacion Ingenieria
LIMA
PERU

29.0 Filters, sieves and strainers

29.1 Filters

MESH FILTER

This equipment is suitable for oil, water etc., and comprises a filter and a sieve. The filter is carbon steel and the sieve is made from stainless steel. The unit is heat-resistant up to 200°C and has a maximum working pressure of 2942kPa (427psi). The mesh has a filtering surface of 40cm².

PRICE CODE: 1

ECICASA INDUSTRIA E COMERCIO LTDA
Rua Guaranesia 900
SAO PAULO
BRAZIL

29.2 Filter presses

AGRICO FILTER PRESS

The liquid to be filtered is pumped into the press and fills the spaces between each set of filter cloths. It then passes through the cloths to the outlet at the bottom of the plate. Solids are retained on the filter cloths. A tray is provided to collect the liquid. Models are available with 16, 18 or 24 plates.

PRICE CODE: 3

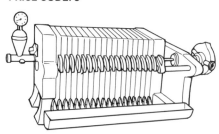

AGRICULTURAL ENGINEERS LIMITED
Ring Road West Industrial Area
P.O. Box 12127
ACCRA-NORTH
GHANA

FILTER PRESS

This press is designed for the purification of oil and consists of a series of iron plates with fabric filters mounted on an iron base. Throughput: up to 600 litres per hour. Two sizes are available: 43 × 43cm, or 58 × 58cm.

PRICE CODE: 4

LAXMI BRASS AND IRON WORKS
Opp. Patel Industrial Estate, Yumana
Mill Road, Pratapnagar
BARODA 390 004
INDIA

STAINLESS STEEL, HORIZONTAL PLATE PRESS

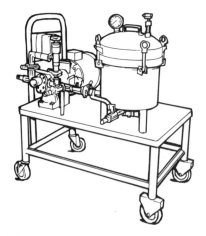

This press filters beverages to clarify them. It comprises a stainless steel tank with a cover containing the filtering unit, which is coupled to a stainless steel pump and motor. The filtering unit consists of a series of disc-type filter plates, perforated screens, filter media and interlocking cups. Feed liquid reaches the plates via several circular openings and suspended solids are retained by the filter media. Finally, the clear liquid runs down the central channel

formed by the interlocking cups and leaves the machine via an outlet pipe. Throughput: 400 litres per hour. Two sizes are available (36 × 20cm, or 46 × 30cm).

PRICE CODE: 4

MASTER MECHANICAL WORKS PVT LTD
Pushpanjali SV Road
Santa Cruz (W)
BOMBAY 400 054
INDIA

FILTER PRESS

This press is fabricated from cast iron and is designed to have a loading of 6kg/cm² working area for filtering oil. The filter plates are truncated square pyramids, arranged in evenly spaced lines. Thus, only a small area of filter cloth is in contact with the metal and a greater proportion is free for filtration. An electrically-driven plunger pump is used for pumping oil to the press.

PRICE CODE: 4

BHARAT INDUSTRIAL CORPORATION
Petit Compound, Nana Chowk
Grant Road
BOMBAY 400 007
INDIA

FILTER PRESS

This press is designed for oil clarification. All chambers are lined with filter cloth and all plates have intersecting shallow channels over their surface with a feed hole in their centre. It operates in a similar way to the presses described above.

PRICE CODE: 4

AULIA AGRO EQUIP INDUSTRIES PVT LTD
36-A Lawerence Road
LAHORE
PAKISTAN

29.3 Sieves

FRUIT SIEVE

This hand-operated sieve will separate fruit pulp and juice.

PRICE CODE: 1

GROUPE DE RECHERCHE ET D'ECHANGES TECHNOLOGIQUES
213 Rue Lafayette
75010 PARIS
FRANCE

MINI-SIFTER

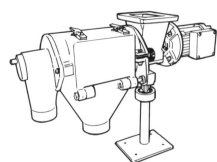

This sieving machine is used for small-scale production. The drive to the machine is fixed and the casing has an inspection door. Throughput: up to 150kg per hour.

PRICE CODE: 2

KEMUTEC GROUP LTD
Hulley Road
Hurdsfield Industrial Estate
MACCLESFIELD
SK10 2ND
UK

SIEVING MACHINE

This simple sieving machine is designed for use with cassava flour. The cassava is fed into the machine from a hopper. The sieved flour is then ready for roasting. The machine has a 0.75kW (1hp) electric motor with a reduction unit.
PRICE CODE: 1
AGRICULTURAL ENGINEERS LTD
Ring Road West Industrial Area
P.O. Box 12127
ACCRA-NORTH
GHANA

VIBRATING SIEVE

This large sieve is designed for processing any starchy foodstuff. It requires a 0.89kW (1.2hp) motor. Throughput: 300kg per hour.
PRICE CODE: 4
INDUSTRIAS TECHNICAS 'DOLORIER'
Alfredo Mendiola 690
Urbanizacion Ingenieria
LIMA
PERU

AGRICO KERNEL/SHELL SEPERATOR

This separates palm kernels from shells. It comprises a simple cylindrical drum covered with a sieve, and rotates on ball bearings. A mixture of broken shells/kernels are fed into the hopper and the drum is turned at a very low speed. The result is clean kernels and the shells are separated into three compartments. Throughput: 500kg material per hour.
PRICE CODE: 1
AGRICULTURAL ENGINEERS LTD
Ring Road West Industrial Area
P.O. Box 12127
ACCRA-NORTH
GHANA

MINI FLOUR-SIFTING MACHINE

This machine sifts flour using a flow of air. The machine has a 0.37kW (0.5hp) electric motor.
PRICE CODE: 2
BIJOY ENGINEERS
Mini Industrial Estate
P.O. Arimpur
Trichur District
KERALA 680 620
INDIA

ROTATING SIEVE (PLANSIFIER)

This sieve separates rice into grades for polishing, using a vibrating round sieve which causes the finer and heavier particles to separate. Single- or double-decked versions are available. Depending on the

model, the machine requires 0.75kW to 1.5kW (1–2hp) power. Throughput: Single: 1500kg per hour, Double: 2200kg per hour.
PRICE CODE: 2
DEVRAJ & COMPANY
Krishan Sudama Marg
Firozpur City
PUNJAB
INDIA

ROTARY FLAT SIEVE

This separates the meal/broken grains from rice after pearling. The rotating sieve sets up a circular eddying motion in the stock which causes the finer and heavier particles to gravitate to the sifting surfaces, leaving the coarser and lighter particles on the top layer. Depending on the model, it requires 0.75kW to 1.5kW (1–1.5hp) power. Throughput: 1500–2500 kg per hour. Two sizes are available: 100 × 200cm and 150 × 200cm.
PRICE CODE: 2
DANDEKAR MACHINE WORKS LTD
Dandekarwadi, Bhiwandi
District Thane
Maharashtra 421 302
INDIA

REFINING MACHINE S40

This machine is used to refine flour . It is fabricated from cast iron and aluminium. Power requirements: 3hp, 60Hz and 3 phase supply. Capacity: 40kg dough. Dimensions: 120 × 100 × 15cm.
PRICE CODE: 4
MAQUINARIA OVERENA S.A DE C.V
Av Division del Norte 2894
Col Parque San Andres
MEXICO DF 04040
MEXICO

IMPROVED PULVERISER AND SIFTER

This is a prototype gari sieve which cuts up the cake into small pieces which then pass on to a conveyor belt to a vibrating sieve. Throughput: 125kg gari per hour.
PRICE CODE: 3
AGRICULTURE ENGINEERING DEPARTMENT
University of Ife
NIGERIA

29.4 Strainers

A common operation which often causes problems is the sieving/straining of wet products. This can be carried out in a pulper/finisher machine, or on a small scale, hand strainers can be used. Hand strainers are effective provided they are made to vibrate, which prevents blockage of the mesh. The following is a simple system that has been used with some success.

A large plastic or stainless steel sieve is hand held against the rotating shaft of a power drill or small electric motor. The shaft end is fitted with an eccentric cam which imparts a rapid vibration to the sieve.

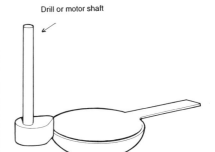
Drill or motor shaft

JELLY BAG

This bag is used to filter juice, jelly and syrups in order to make crystal clear products. It is vital that the bag should be clean and it should be thoroughly washed and boiled after each use.
PRICE CODE: 1
BRITAM
5 Ferry Lane Industrial Estate
BRENTFORD
TW8 0AW
UK

JELLY STRAINER

This will strain jelly, juices, saps, wine, beer etc. It is free-standing with a strainer bag, and will fold flat for storage. The bag has a capacity of 2kg. Height: 51cm.

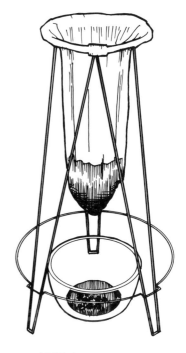

PRICE CODE: 1

JELLY STRAINER BAG

Designed for the straining of juices etc. this has a bag which acts as a filter. In addition, it can be used in conjunction with the jelly-strainer stand. Capacity: 2kg. Dimensions: diameter 15cm × length 30cm.
PRICE CODE: 1
LAKELAND PLASTICS
38 Alexandra Buildings
WINDERMERE LA23 1BQ
UK

CONICAL STRAINER

This honey strainer has a fine grading. It is made from stainless steel and is fitted with a rim. A round strainer is also available.

PRICE CODE: 1

SWIENTY
Hortoftvej 16
Ragebol
Sonderborg
DK 6400
DENMARK

CONICAL TAP STRAINER

This honey strainer is made from stainless steel. It operates by attachment to the extractor tap for the initial coarse straining. Other versions include a double-slide strainer. Diameter: 13cm.

PRICE CODE: 1

STEELE & BROODIE
Stevens Drove
Houghton
STOCKBRIDGE
Hampshire
UK

UNCAPPING CAN

This drains the honey from cappings. While the honey drains through and is removed from the base of the can by the honey gate, the capings fall onto the strainer. The can is made from 61cm gauge, heavy tinned steel. Dimensions: 77.5 × 40.6cm.

PRICE CODE: 1

PENDER BEEKEEPING SUPPLIES PTY LTD
17 Gardiner Street
Rutherford
NSW
2320
AUSTRALIA

FILTER SOCKS

These are used to filter milk. Various brands are available: Maxiflow, Blueline, Redline and Flowmade. All brands are packed in boxes of one hundred.

PRICE CODE: 1

FILTER MASTER MILK FILTER

This filters and strains milk.

PRICE CODE: 1

UNIFLOW STRAINER

This is capable of straining up to 18 litres of milk at a time.

PRICE CODE: 1

MILK STRAINER

This milk strainer is made from anodized aluminium. A heavy gauge stainless steel strainer is also available.

PRICE CODE: 1

HOMILEAF MILK STRAINER

This aluminium hand-operated milk strainer has a total capacity of 6.8 litres.

PRICE CODE: 1

STAINLESS STEEL SIEVE

This sieve is used primarily for straining cheese, but it can also be used for vegetables. It has a diameter of 18cm.

PRICE CODE: 1

PERFORATED FLAT LADLE

This stainless steel ladle is used for draining or straining cheeses. It has a diameter of 10cm.

PRICE CODE: 1

All the above strainers are available from:

R J FULLWOOD & BLAND
ELLESMERE
Shropshire
SY12 9DF
UK

30.0 Flaking and splitting machines

DAL SPLITTER

This machine splits shelled pulses into dal. Its power requirements are 0.75kW (1hp). Throughput: 1000kg per hour.

PRICE CODE: 3

DANDEKAR MACHINE WORKS LTD.
Dandekarwadi, Bhiwandi
District Thane
Maharashtra
421 302
INDIA

Rice flaking equipment

LOW-COST ROLLER FLAKING MACHINE

This portable machine is used for both cereals and pulses. It consists of a hopper, hollow rollers, gear transmission system and support frame. The rollers are fabricated from mild steel pipe and are surface plated with nickel. The power requirements are a 0.75kW (1hp) single phase electric motor through a belt drive to the central roller. Throughput: 20kg per hour soybean or 10kg per hour rice. Weight: 80kg.

PRICE CODE: 2

CENTRAL INSTITUTE OF AGRICULTURE ENGINEERING
Nabi Bagh
Bersala Road
BHOPAL
462018
INDIA

RICE FLAKE MACHINE (POHA MILL)

This power-driven mill has a 28cm roller and a 76cm drum, both made of cast iron. 60–70kg of paddy can be made into poha per hour. The mill is free-standing. The roller revolves while the paddy is being pressed and can be adjusted according to the thickness of poha required. By pressing the roller against the rim of the drum, all poha can be collected near the centre of the drum.

PRICE CODE: 3

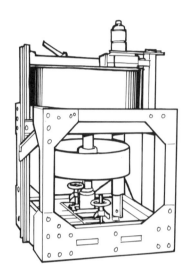

DANDEKAR BROTHERS (ENGINEERS & CO)
Shivaji Nagar, Factory Area
SANGLI 416 416
INDIA

Rice flaking equipment also manufactured by:

DANDEKAR MACHINE WORKS LTD.
Dandekarwadi, Bhiwandi
District Thane
Maharashtra
421 302
INDIA

BEHERE'S & UNION INDUSTRIAL WORKS
Jeevan Prakash
Masoli
DAHANU ROAD 401 602
District Thane
INDIA

 # 31.0 Flow meters

These are more commonly found in larger-scale continuous processes where they are used to measure the flow of material through the process or the level of foods on storage tanks. There are a large number of different designs of varying complexity and cost. Simple flow meters may find application at the larger range of small-scale processing operations.

SERIES 1100 FLOW METERS

This flow meter has a steel channel at each end and a connection block containing the fittings which secure the tube. The block at the inlet end may also incorporate a needle valve. The front of the instrument is protected by a transparent plastic dust-cover. Available in a range of tube lengths (all calibrated). They have a working temperature of 60°C maximum.

PRICE CODE: 2

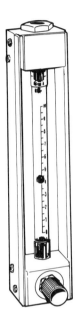

FISHER CONTROLS
Rotameter Works
330 Purley Way
CROYDON
UK

TROGAMID FLOW METERS

Suitable for measuring the flow of almost any liquid or gas.

PRICE CODE: 2

NIXON INSTRUMENTATION
Charlton Kings Industrial Estate
Cirencester Road
CHELTENHAM
UK

Flow meters also manufactured by:

PLATON FLOWBITS
Wella Road
BASINGSTOKE
UK

DANFLOSS LTD
Instrumentation Division
Magflo House
Ebley Road
STONEHOUSE
UK

 32.0 Freezers

NON-ELECTRIC CHEST FREEZER

This freezer uses liquid petroleum gas, natural gas or kerosene. The unit has a child-proof key lock and tight magnetic gaskets on the door. A parts kit is also available. Dimensions: height 102cm × weight 95cm × depth 77cm.

PRICE CODE: 4

LEHMAN HARDWARE & APPLIANCES INC.
P.O. Box 41
4779 Kidron Road
Kidron
OHIO 44636
USA

 33.0 Fryers

Fryers (gas and electric)

Country	Manufacturer	Equipment name	Capacity (litres of oil)	Price code	Power (kW)	Other information
Brazil	Braxinox Brazil Equip Industr Ltda	INDUSTRIAL FRYER ETE MM 220	40	4	36	
	Saveiro Comercio e Industria	FRYER	9	2		
	Croydon Ind de Mags Ltda	FRYER MODEL FA2.5	27	3	5	
		MODEL F2C		3	5-6	
		MODEL TFSL	10	2	2.5-3	
		MODEL TF GLP	11.5	2		Uses gas supply
		MODEL F1	12	2	2.5-3	Gas model available
		F2-5/F2-8	12	2	5	Gas model available
		F2GS	12	2		Uses gas supply
		SAUSAGE HOT FRYER FG1		2		Uses gas supply
		SAUSAGE FRYER CH1		3		Temperature range: 50-300°C
India	Continental Equipments Pvt Ltd	DEEP FAT FRYER	8	2	3	
	Gardeners Corporation	DEEP FAT FRYER		3		Gas and electric fryers available
	Thakar Equipment Co.	ELECTRIC FRYER	30	3	9	
		MODEL ESF30		3	12	
		MODEL EDF1		3		Other sizes available
		GAS FRYER	60	3		
Nigeria	Department of Agricultural Eng.	GARI FRYING MACHINE		3		Fries 66kg gari/h
Philippines	Somerville Stainless Steel Corp.	FAT FRYER		2		Other sizes available
UK	Olivertons	M LINE ELECTRIC/GAS FRYERS		3		Two electric and two gas models are available
		FRYMASTER RANGE		3		Uses gas supply. BABY, MAJOR and STANDARD are available

Fryers (continued)

Country	Manufacturer	Equipment name	Capacity (litres of oil)	Price code	Power (kW)	Other information
Zimbabwe	GDI (Pvt) Ltd	CHIP FRYERS		2		SINGLE or DOUBLE model
	Mitchell and Johnson PVT Ltd	DEEP FRYER	12	2	6	
	Plaza Oriental (PVT) Ltd	WOK		1		Flat or rounded models available
	Stainless Steel Industries Pvt Ltd	CHIP FRYER		2	4	

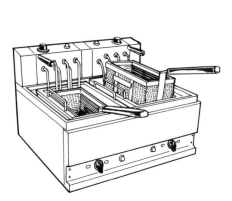

Deep fat fryers

34.0 Grating equipment

34.1 Manual grating equipment

PEDAL-OPERATED COCONUT GRATER

This portable grater can be operated by two people. The rotary blades (which are detachable), are mounted onto a bicycle. While one operator presses half of the coconut against the blades, the other turns the pedals. The grater blades can be subsituted for slicing, milling, shelling and winnowing operations.

PRICE CODE: 1

DEPT OF AGRICULTURE, CHEMISTRY & FOOD SCIENCE
Visayas State
College of Agriculture
Baybay, LEYTE 7127-A
PHILIPPINES

ROOT CROP GRATER/PULVERIZER

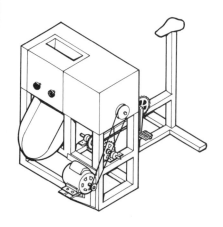

This machine is used to produce grated root crops for processing to various food

products. The pulverizer can use either pedal or motor power. There is an easy-to-clean grating drum, power transmission, hopper and an aluminium trough. Throughput: 50kg per hour.

PRICE CODE: 1

VISAYAS STATE COLLEGE OF AGRICULTURE
8 Lourdes st
PASAY CITY 3129
PHILIPPINES

Grating equipment also manufactured by:

LAKELAND PLASTICS
38 Alexandra Buildings
WINDERMERE
LA23 1BQ
UK

ILO – SDSR
P.O. Box 60598
NAIROBI
KENYA

34.2 Powered grating equipment

WADHWA CASSAVA GRATER

This machine, powered by a 3.7kW (5hp) diesel engine or electric motor has circular grating discs. The discs run at 1550 rpm and are self-cleaning. Throughput: 500kg cassava per hour.

PRICE CODE: 2

AGRICULTURAL ENGINEERS LTD
Ring Road West Industrial Area
P.O. Box 12127
ACCRA-NORTH
GHANA

CASSAVA GRATER

Power: 1.5kW (2hp) or 2.2kW (3hp) or 3.7kW (5hp), depending on model. Throughput: 200–300kg per hour.

PRICE CODE: 2–3

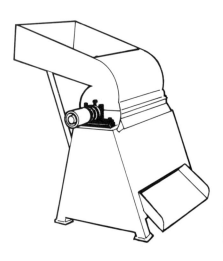

RAJAN UNIVERSAL EXPORTS (MFRS) LTD
Raj Buildings 162, Linghi
Chetty Street, P. Bag No. 250
MADRAS 600 001
INDIA

VERTICAL DRUM GRATER

Root crops such as cassava can be grated with this machine. The surface of the drum is covered with a sheet of perforated metal which rotates across the base of the feed hopper by a belt drive from a 2.9kW (4hp) diesel or electric motor. The cassava is pressed against the grating surface by a wooden block inserted into the hopper. The hopper, drum and frame can be made from timber or scrap metal.

PRICE CODE: 1

TIKONKO AG. EXTENSION CENTRE
Methodist Church Headquarters
Wesley House
P.O. Box 64
FREETOWN
SIERRA LEONE

Powered grating equipment also manufactured by:
INDUSTRIAS MAQUINA D'ANDREA S.A.
Rue General Jardim 645
SAO PAULO 01223
BRAZIL

INSTITUTE OF SCIENCE & TECHNOLOGY
University of the Philippines
Los Banos
LAGUNA
PHILIPPINES

35.0 Grills

PORTABLE FOOD GRILL

This grill is suitable for use both inside and outside. Charcoal is used, and there is a heat regulator and liquid gas/alcohol reservoir.

PRICE CODE: 1

DARMO METAL INDUSTRIES
Tempaurel Compound
BO Putatan, Muntinlupa
METRO MANILA
PHILIPPINES

Grills also manufactured by:
THAKAR EQUIPMENT CO.
66 Okhla Industrial Estate
NEW DELHI 110 020
INDIA

36.0 Heaters and hotplates

CHALLENGER WATER HEATERS

These models can be used for a continuous supply of boiling water. The urns are lined with copper for maximum heat transfer and all pipework is contained within a stainless steel cover. Optional extras include a draw-off swivel arm, drip tray and boiler stand. Three models are available, with throughputs of 142–255 litres hour.

PRICE CODE: 2–3

BULK BOILING URNS

These stainless steel urns provide immediate hot water. There is a 'low-water' safety cut-out mechanism and a selector switch. Optional extras include: draw-off tap, swivel arm, drip tray and boiler stand. The urns are available as either gas or electric powered, and there are five sizes from 18–54 litres.

PRICE CODE: 2–3

DIGITHERM

This electric tilting kettle has an insulated exterior surface which cannot exceed 40°C.

A thermostat ensures an even heat distribution, and prevents adhesion and boiling. There are four individually controlled heating elements. Other models available include: 40, 60, 80 and 100l. Capacity: 20l.

PRICE CODE: 2–4

All the above heaters are available from:
STOTT BENHAM LTD
P.O. Box 9, High Barn Street
Vernon Works
Royton
OLDHAM
OL7. 6RP
UK

Water heaters also manufactured by:
THAKAR EQUIPMENT COMPANY
66 Okhla Industrial Estate
NEW DELHI 110 020
INDIA

GAS HOTPLATES

These are designed for small-scale heating. Each hotplate consists of cast iron burners with holes and air mixers. The burners are easy to convert from liquid propane to natural gas. In addition, each burner is adjustable from a low to a simmer flame. The handles are made from bakelite to prevent burning. Versions available (i.e. no. of burners): 1, 2, 3.

PRICE CODE: 1

LEHMAN HARDWARE & APPLIANCES INC.
P.O.Box 41
4779 Kidron Road
Kidron
OHIO 44636
USA

OLD ROMAN PIZZA STOVE

Pizza crusts can be made using this stove. The pizza mix is poured onto the stove, spread and heated.

PRICE CODE: 1

LEHMAN HARDWARE & APPLIANCES INC.
P.O. Box 41
4779 Kidron Road
Kidron
OHIO 44636
USA

KRUMKAKE BAKER

This hotplate is designed for making thin biscuits. A wooden cone roller is provided to roll the biscuit into a cone.

PRICE CODE: 1

VITANTONIO
Dept C
34355 Vokes Dr
Eastlake
OHIO 44094
USA

HEAT EXCHANGER

This consists of plates which form a pair of concentric loops at even spaces. The edges of the loops are sealed so that liquid can flow through the channels. Models are available for an exchange temperature of up to 600°C.

PRICE CODE: 2

LARSEN & TOUBRO LTD
L & T House, Ballard Estate
P B No 278
BOMBAY 400 038
INDIA

ELECTRIC BAIN-MARIE

This can be used as a crockery warmer or as a service counter. Dimensions: length 4.27m × width 1.07m.

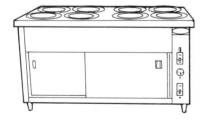

PRICE CODE: 2

THAKAR EQUIPMENT COMPANY
66 Okhla Industrial Estate
NEW DELHI 110 020
INDIA

COCINAS MODEL 2MMP

This kerosene stove is suitable for use in a small industrial setting. Two gas burners are encased in a cabinet with a flat heating surface. Many variations of this model are available.

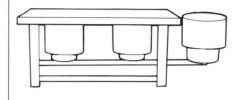

PRICE CODE: 1

CASTROVIRREYNA SRL
Jr Catsrovirreyna 876
Casila Postal 050033 Brena
LIMA
PERU

GRIDDLES

The following sizes are available:
(a) 24 × 18 × 15cm **PRICE CODE: 1**
(b) 24 × 24 × 15cm **PRICE CODE: 2**
(c) 6 × 24 × 15cm **PRICE CODE: 2**

SOMERVILLE STAINLESS STEEL CORP.
90 Sen Gil J Pujat Ave
Makati
METRO MANILA
PHILIPPINES

ANAGI II

This ceramic stove is designed to burn all woods including rubber wood.

It has two pot positions and one fire box. The fuel is fed in at the side of the larger pot hole and both cooking positions are immediately in front of the cook. A connecting tunnel between the firebox and the second pot hole distributes approximately one third of the heat output which is sufficient to maintain gentle boiling. The fuel efficiency of this stove is approximately 20%. Construction details may be obtained from:

INTERMEDIATE TECHNOLOGY
DEVELOPMENT GROUP
Myson House
Railway Terrace
RUGBY CV21 3HT
UK

Hot plates also manufactured by:

ESSCO FURNACES (P) LTD
127 Velachery Road
Little Mount, Saidapet
MADRAS 600 015
INDIA

THAKAR EQUIPMENT COMPANY
66 Okhla Industrial Estate
NEW DELHI 110 020
INDIA

CROYDON IND DE MAQS LTDA
Soci Gerente
R 24 de Fevereio 71 F75-77-79
RIO DE JANEIRO
RJ 21040
BRAZIL

Crêpe and waffle makers

CRÊPE MAKING MACHINE

This machine makes crêpes suitable for the catering industry. It is made of fibreglass and has an injector pump for dough and two grills. Power requirements are 110/220V. Throughput: 200 × 40g crêpes per hour.

PRICE CODE: 2

MARQUINAS DE KREPS SUISSO
Rua Oscar Freidre 1784
01426 SAO PAULO SP
BRAZIL

WAFFLE IRON

This is designed for use on the top of a cooking stove. It has a non-stick teflon interior, heatproof bakelite handles, and a thermometer to ensure even baking. The batter is placed inside and the top is pressed down. Four waffles can be made per batch.

PRICE CODE: 1

LEHMAN HARDWARE & APPLIANCES INC.
P.O. Box 41
4779 Kidron Road
Kidron
OHIO 44636
USA

Sandwich and waffle makers are also manufactured by:

CROYDON IND DE MAQS LTDA
Soci Gerente
R 24 de Fevereio 71 F75-77-79
RIO DE JANEIRO
RJ 21040
BRAZIL

Rice cookers

RICE COOKER

This automatic rice cooker has a power requirement of 0.35–1.5kW, depending on the model. Capacity 0.6–4.2 litres. Various other sizes are available.

PRICE CODE: 1

GUANGZHOU ELECTRIC RICE COOKER FACTORY
Lianhe Suburbs
NORTH GUANGZHOU
CHINA

Rice cookers also manufactured by:

MR FABRICATION AND SERVICING CORP.
Carmelo & Bauermann Compound
EDSA, Guadalupe
Makati
METRO MANILA
PHILIPPINES

Food warmers

ELECTRIC HOT CUPBOARD

This keeps food hot (above 60°C) to prevent the growth of food poisoning bacteria during storage prior to serving, or it can be used to warm up food.

Three sizes are available:

MODEL	SIZE	POWER
HCE-1	122 × 76cm	1.0kW
HCE-2	183 × 76cm	1.5kW
HCE-3	244 × 76cm	3.0kW

PRICE CODE: 1

THAKAR EQUIPMENT CO.
66 Okhla Industrial Estate
NEW DELHI 110 020
INDIA

FOOD WARMER

This has a removable stainless steel crumb tray with three slatted stainless steel shelves. The front is fitted with heat resistant glass and there is a glass sliding door at the rear.

PRICE CODE: 3

MITCHELL AND JOHNSON PVT LTD
P.O. Box 966
BULAWAYO
ZIMBABWE

Food warmers also manufactured by:

BRAXINOX BRASIL EQUIPAMENTOS INDUSTRIAIS LTDA
Estrada da Servido 155 OSASCO
Caixa Postal 280
06170 SAO PAULO SP
BRAZIL

GRUENTHAL AND BEKKER (PVT) LTD
P.O. Box 2550
HARARE
ZIMBABWE

Gas ring burners

THAKAR BURNERS

MODEL GLR 2
This is a two-burner, gas cooking-range. Several other models (three- and four-burner versions) are also available.

PRICE CODE: 1

THAKAR EQUIPMENT CO.
66 Okhla Industrial Estate
NEW DELHI 110 020
INDIA

37.0 Homogenizers

Homogenizers are used in the food industry to form a stable emulsion from two immiscible liquids, usually oil and water. Homogenization causes the breakdown and dispersal of fat globules to form an homogeneous liquid. In addition, it changes the functional properties or eating quality of foods, although it has no effect on the nutritional value or shelf-life. Products which are homogenized include dairy products such as milk (e.g. 'homogenized milk'), ice cream, butter and margarine. An emulsifying agent, such as lecithin may be added to assist in the process. All homogenizers work on one of three basic principles: pressure, rapid mixing or ultrasonic vibration. The second type is more likely to be suitable for small-scale operations, but all types of equipment are expensive to buy and maintain. Many different types of equipment are available, ranging from simple hand-operated machines to larger, powered homogenizers with greater throughputs.

MIXER/HOMOGENIZER, SILVERSON L4R

This is a heavy duty, multi-purpose, high-speed mixer. Speed control is by means of an integral solid state electronic regulator incorporating an on/off switch. The mixer is supported on an electrically-operated rise and fall bench stand which is activated by push-button control.

The machine is supplied with an emulsifier head, square hole, high sheer screen, general purpose disintegrator head, axial flow head and a pump head. It requires a 220/240V, 50Hz mains supply. Capacity: 12 litres. Dimensions: 30 × 50 × 93cm.

PRICE CODE: 3

THE NORTHERN MEDIA SUPPLY LTD
Sainsbury Way
HESSLE
HU13 9NX
UK

ELECTRICALLY-OPERATED BLUE CALF

Liquid is forced at a constant rate through a homogenizing jet giving good results on rough emulsions and suspensions, provided the liquid can flow through the pump. A safety valve, air vent and jet control are provided, and the equipment is easy to dismantle for cleaning purposes. All parts are manufactured from stainless steel. Throughput: 45 litres per hour. Larger units are also available.

PRICE CODE: 4

HAND-OPERATED QP

This laboratory homogenizer has stainless steel contact parts. There are also bench clamps and jets for fine and coarse suspensions. The 0.5-litre hopper is emptied every 2–3 minutes.

PRICE CODE: 2

The above are both available from:

ADELPHI MANUFACTURING CO. LTD
Olympus House
Mill Green Road
HAYWARDS HEATH
RH16 1XQ
UK

MINISONIC ULTRASONIC HOMOGEN 4065

This is a low-cost, small batch production homogenizer. The stainless steel gear pump takes process material from the vessels and delivers it at a pressure of 1240kPa (180psi) to the ultrasonic homogenizing head.

The homogenizer can be used in several ways:

(1) Rough premix pumped from the vessels through the homogenizing head and discharged into a separate vessel.
(2) Emulsion traverses from one vessel to another and back again.
(3) Separate phases of the emulsion are put into the two vessels. The flow is controlled by a three-way valve.
(4) The continuous phase is recycled through the homogenizer, while the disperse phase is introduced at a controlled rate, via the three-way valve.

Requires 220V, 50–60Hz electricity supply. Motor power: 0.37kW (0.5hp). Throughput: 120 litres per hour. Dimensions: length 50 × width 26 × height 50cm.

PRICE CODE: 4
LUCAS DAWE ULTRASONICS
Concord Road
Western Avenue
LONDON W3 OSD
UK

BUTTER WORKER

This hand-operated butter worker, mixes and works butter ready for forming.

PRICE CODE: 1
DAIRY UDYOG
C-229A/230A, Ghatkopar Ind Est
L.B.S. Marg, Ghatkopar (W)
BOMBAY 400 086
INDIA

EMULSIFIERS/HOMOGENISER HSE 2/20

This machine operates in an up/down or sideways movement, and is available with a stand or a wall support. Capacity: 2–10 litres. Other sizes are available.

PRICE CODE: 2
MAMKO PROJECT ENGINEERING & CONSULTANCY
Yashodham Office Complex
Gen. Arunnkumar Vaidya Marg
Goregon(E)
BOMBAY 400 063
INDIA

HIGH-SPEED EMULSIFIER

This consists of a stainless steel perforated screen disc-impeller with a 0.19kW (0.25hp) motor.

PRICE CODE: 2
NARANGS CORPORATION
25/90 Connaught Place
Below Madras Hotel
NEW DELHI 110 001
INDIA

Homogenizers also manufactured by:

LAKSHMI MILK TESTING MACHINERY CO.
A90 Group Industrial Area
Wazirupa
NEW DELHI 110 052
INDIA

JANKLE & KUNKEL (UK) LTD
P.O. Box 16
LEWES
BN7 3LR
UK

38.0 Ice cream making equipment

Manual ice cream makers

ICE CUBE TRAY

Small ice cubes can be made using these plastic trays which will not break or crack with the cold of the freezer.

PRICE CODE: 1

LOLLY MOULD

This comprises a mould and re-usable sticks for making up to six ice lollies at a time.

PRICE CODE: 1
LAKELAND PLASTICS
38 Alexandra Buildings
WINDERMERE
LA23 1BQ
UK

ICE CREAM FREEZER

This is a large hand-operated ice cream maker. It has cast iron mixers which are hot-dipped in tin to prevent rusting. The scrapers are made from hardwood and the machine has an extra thick tub for fast freezing. Capacity: 22.7 litres ice cream. Height: 61cm.

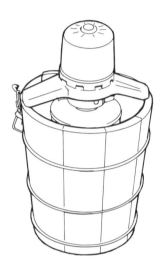

PRICE CODE: 3
LEHMAN HARDWARE & APPLIANCES INC.
P.O. Box 41
4779 Kidron Road
Kidron
OHIO 44636
USA

HOME ICE CREAM MAKER

A handle is turned to operate the machine and up to six servings can be made at once.

PRICE CODE: 1
R J FULLWOOD & BLAND LTD
ELLESMERE
Shropshire
SY12 9DF
UK

ICE CREAM BEATER

Designed for the small scale, this manual ice cream beater has a lever arm which triggers a swivelling and up-and-down motion of the ice cream in the pan. The pan's base rotates, and the mixture is then blended. Input capacity: 9.1 litres. Dimensions: height 167.6cm × diameter 45.7cm.

PRICE CODE: 1
CORNELIO CARREON
9 Figeuroa St
Cebu City
CEBU
PHILIPPINES

Powered ice cream makers

ICE CRUSHER

This crusher has a 3kW (4hp) motor. Overall dimensions: 54 × 32 × 22cm.

PRICE CODE: 3
CROYDON INDUSTRIA DE MAQINAS LTDA
Soci Gerente
R 24 de Fevereio 71 F75-77-79
RIO DE JANEIRO, RJ 21040
BRAZIL

HOME ICE CREAM MAKER

With this ice cream maker approximately 1.2 litres of ice cream can be made in 25–40 minutes.

PRICE CODE: 1
R J FULLWOOD & BLAND LTD
ELLESMERE
Shropshire
SY12 9DF
UK

WHITE MOUNTAIN ICE CREAM FREEZERS

This machine can be hand- or electrically-operated for small-scale ice cream production. The three-gear drive mixes and beats the entire contents of a wooden tub. The motor requires 115V AC and turns at 1200 rpm. Several sizes are available.

PRICE CODE: 2
LEHMAN HARDWARE & APPLIANCES INC.
P.O. Box 41
4779 Kidron Road
Kidron
OHIO 44636
USA

Powered ice cream making equipment also manufactured by:
TAYLOR FREEZER UK PLC
Denmark House
Old Bath Road
TWYFORD
RG10 9QJ
UK

38.1 Ice cream scoops

ICE CREAM SCOOP

A plastic version.

PRICE CODE: 1
R J FULLWOOD & BLAND LTD
ELLESMERE
Shropshire
SY12 9DF
UK

BONZER 'EEZISCOOP' PORTIONER

This is used for portioning ice cream and similar foods. The design has a moulded bowl.

PRICE CODE: 1

BONZER LITEGRIP PORTIONER

This is used for portioning ice cream and similar foods. It is available in eight metric sizes, and the design includes lightweight arms with built in springs for ease of operation.

PRICE CODE: 1
MITCHELL & COOPER LTD
Framfield Road
UCKFIELD
TN22 5AO2
UK

ANTIFREEZE ICE CREAM SCOOP

This small hand-operated scoop, has antifreeze sealed inside. Length: 17.8cm.

PRICE CODE: 1

ICE CREAM SCOOP

This hand-operated scoop is made of stainless steel, length 21cm.

PRICE CODE: 1
LEHMAN HARDWARE & APPLIANCES INC.
P.O. Box 41
4779 Kidron Road
Kidron
OHIO 44636
USA

Scoops also manufactured by:
LAKELAND PLASTICS
38 Alexandra Buildings
WINDERMERE
LA23 1BQ
UK

M J & D PRICE
Leworthy Mill
WOOLFARDISWORTHY
EX39 5PY
UK

39.0 Incubators

This equipment comprises insulated, heated boxes which are used to hold food at a preset temperature, usually to allow micro-organisms to grow (e.g. incubation of dough and yoghurts).

COMMERCIAL YOGHURT MAKER

This is an electric incubator used for small- to medium-scale yoghurt production. Two sizes are available:
1. The Y140, with a capacity of 140 pots/cycle.
2. The Y300 with a capacity of 300 pots/cycle.

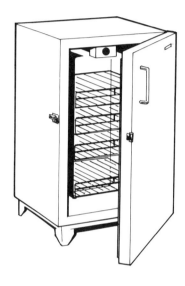

PRICE CODE: 3

BEL NATURAL YOGHURT MAKER

This is designed for small-scale yoghurt production and does not require electricity. A thermometer is included. Capacity: one litre.

PRICE CODE: 1
R J FULLWOOD & BLAND LTD
Ellesmere
Shropshire
SY12 9DF
UK

CHEESE AND YOGHURT INCUBATOR

This incubator distributes heat evenly to yoghurt and cheese cultures.

PRICE CODE: 1

LEHMAN HARDWARE & APPLIANCES INC.
P.O. Box 41
4779 Kidron Road
Kidron
OHIO 44636
USA

Incubators are also manufactured by:

LAKELAND PLASTICS
38 Alexandra Buildings
WINDERMERE LA23 1BQ
UK

GEBR RADEMAKER
P.O. Box 81
3640 AB MIJDECHT
NETHERLANDS

RR MILL INC.
45 West First North
Smithfield
UTAH
84335
USA

40.0 Labelling machines

Manual labelling machines

LABEL GUMMING MACHINE

This hand-operated, label gumming machine is suitable for labels of up to 15cm width.

PRICE CODE: 1

NARANGS CORPORATION
25/90 Connaught Place
Below Madras Hotel
NEW DELHI 110 001
INDIA

LABEL GUMMING MACHINE

A pack of labels is placed at the feeding end of the machine and are automatically fed through. When the labels reach the other end they are uniformly gummed, and are ready to be picked up by the packers. Requires 230V AC. Throughput: 20–40 labels per operation. A hand-operated model is also available.

PRICE CODE: 2

J T JAGTIANI
National House, Tulloch Road
Apollo Bunder
BOMBAY 400 039
INDIA

SEMI-AUTOMATIC GUMMING AND LABELLING MACHINE

This machine is designed for labelling onto round and flat surfaces. Depending on the size, the label can cover up to 50% of the circumference of a round bottle. Bottles are placed in the machine and taken out manually. Motor power: 0.37kW (0.5hp). Throughput: 1800 bottles per hour. Dimensions: Length 45.7 × Width 61 × Height 152.4cm.

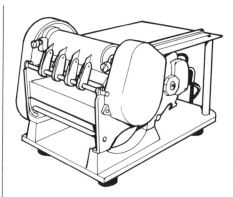

PRICE CODE: 2

In addition, another gumming/labelling machine is available with a range of roller sizes from 12.7cm to 45.7cm in length.

MASTER MECHANICAL WORKS PVT LTD
Pushpanjali SV Road
Santa Cruz (W)
BOMBAY 400 054
INDIA

LABELLER/GLUER

This all purpose gluer has brass rollers, guides, feed plate, and a corrosion-proof glue pot. The unit can use several sizes and shapes of labels, and uses a 0.02kW (0.03hp) motor. Dimensions: 42.5 × 24.1 × 14cm.

PRICE CODE: 2

FMC CORPORATION
Machinery International Division
P.O. Box 1178
SAN JOSE
California 95108
USA

41.0 Mills and grinders

41.1 Plate mills

No. 2 CAST HAND MILL

This is a small hand-operated mill suitable for all dry materials. It has a counter-balanced crank and cone shaped burrs. Throughput: 11.3kg per hour. Length: 40.6cm.

PRICE CODE: 1

LEHMAN HARDWARE & APPLIANCES INC.
P.O. Box 41
4779 Kidron Road
Kidron
OHIO 44636
USA

THE 'LITTLE ARK' GRAIN MILL

This hand-operated small-scale milling machine is convertible from stones to burrs. Burrs are useful for pulses and beans, and stones are useful for flours. The mill can be safely motorized as it has bronze brushes on the drive shaft. It stands 37.5cm high.

PRICE CODE: 1

LEHMAN HARDWARE & APPLIANCES INC.
P.O. Box 41
4779 Kidron Road
Kidron
OHIO 44636
USA

GRINDING MILL

This hand-grinding mill has adjustments for fine and coarse grinding. It is fabricated from metal and is powered by a large hand wheel.

PRICE CODE: 2

G NORTH & SON (PVT) LTD
P.O. Box BPX ST 111
Southerton
HARARE
ZIMBABWE

GRAIN GRINDING MILL

This is an all-metal hand-powered maize grinding mill. It has a simple adjustment for fine or coarse meal grinding and can also kibble grain for poultry. The mill is mounted on a cast iron pedestal and is turned by a handle on a large iron flywheel. Weight: 101kg.

PRICE CODE: 2

G NORTH & SON (PVT) LTD
P.O. Box BPX ST 111
Southerton
HARARE
ZIMBABWE

HAND FLOUR MILL

Designed for use with cereals, this bench mounted mill has a small hopper, handle and flywheel. Throughput: 5–50kg per hour.

PRICE CODE: 2

CECOCO
P.O. Box 8
Ibaraki City
OSAKA 567
JAPAN

HAND FLOUR MILL

This mill is used for cereals, although it uses meat chopper parts. Throughput is 6kg per hour.

PRICE CODE: 1

CECOCO
P.O. Box 8
Ibaraki City
OSAKA 567
JAPAN

Manual plate mills also manufactured by:

AGRO MACHINERY LTD
P.O. Box 3281
Bush Rod Island
Monrovia
LIBERIA

INSTITUTE FOR AGRICULTURAL RESEARCH
Samaru, Ahmadu Bello Uni
PmB 1044
ZARIA
NIGERIA

C.S. BELL CO.
170 W Davis Street
Box 291
Tiffin
OHIO 44883
USA

LEHMAN HARDWARE & APPLIANCES INC.
P.O. Box 41
4779 Kidron Road
Kidron
OHIO 44636
USA

RR MILL INC.
45 West First North
Smithfield
UTAH 84335
USA

Powered plate mills

GRINDING MILL

This cast iron mill can be used for wheat, barley, peas, rice and wet/dry maize. The runner stone rotates with the shaft and the fixed stone is attached to the casing towards the pulley side. A safety spring ensures that the stones are protected from hard foreign matter. All parts should be lubricated and the stones should be chiselled once a week. Throughput is 90–260 kg per hour. Versions available include: 12V, 16V, 18V, 20V.

PRICE CODE: 3

SANJAY STEEL BUILDERS
Matru Chhaya N Subbash Road
Near Syndicate Bank, Mulund
BOMBAY 400 080
INDIA

DISC GRINDER

This machine reduces a crushed product to a fine particle size. The machine has two 20.3cm diameter grinding discs: one fixed, and one rotating. The rotating disc is mounted on a horizontal shaft and is driven by a 2.7hp (2kW) motor. The material falls from a feed chute through a central hole in the fixed disc into the grinding chamber and the ground material is discharged from a chute at the front of the machine. Throughput: 50kg per hour.

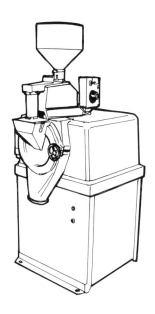

PRICE CODE: 4

THE PASCALL ENGINEERING CO. LTD
Gatwick Rd
CRAWLEY
RH10 2RS
UK

UPLD IMPROVED VILLAGE RICE MILL

Rice can be milled using this machine which is operated by a 16hp (12kW) diesel engine or a 10hp (7.5kW) electric motor. It can be constructed from locally available materials. Throughput: 300kg paddy per hour. Dimensions: length 100 × height 180cm.

PRICE CODE: 3

AGRICULTURAL MECHANIZATION DEVELOPMENT PROGRAMME
AMDP, CEAT, UP Los Banos College
LAGUNA 4031
PHILIPPINES

BEST GRAIN MILL

This small-scale grain mill comes with a balanced V-belt pulley that doubles as a flywheel for hand cranking. It requires 0.25kW (0.3hp) and produces a very fine flour. Throughput: (by hand) 9kg per hour. Dimensions: 35cm to top of hopper.

PRICE CODE: 2

LEHMAN HARDWARE & APPLIANCES INC.
P.O. Box 41
4779 Kidron Road
Kidron
OHIO 44636
USA

COFFEE MILL RA23E

This small coffee mill is equipped with a 0.25kW (0.3hp) single phase motor, and box for the ground coffee.

PRICE CODE: 2

JR ARAUJO IND E COM DE MAQS LTDA
Division Internacional
Rua General Jardim 645
SAO PAULO SP
CEP 01223
BRAZIL

VERTICAL STONE GRINDER

This comprises stone grinders and has heavy dust-proof ball bearings for smooth running. The clearance between the stones can be adjusted easily. It requires a 5–15hp motor depending on the model. Throughput: 500–600kg per hour. Size range: 30, 40.6, 45.7, 50.8cm diameter stones.

PRICE CODE: 4

RAJAN UNIVERSAL EXPORTS (MFRS) P LTD
Raj Buildings 162, Linghi
Chetty Street, P. Bag No. 250
MADRAS 600 001
INDIA

GROUNDNUT GRINDER

This consists of a feed hopper, grinding roller, outlet chute and motor drive, all mounted on a single frame. Throughput: 275kg per hour.

PRICE CODE: 3

KUNASIN MACHINERY
107–108 Sri-Satchanalai Road
Sawanankalok
Sukothai
THAILAND

'MILL RITE' ELECTRIC GRAIN MILL

This self-contained electric mill grinds flours or oily materials such as soy beans, corn etc. The motor has no brushes or slip rings to replace. The gear head is fully enclosed, lubricated and protected from dust. Throughput: 4.5–6.8kg per hour. Dimensions: height 33 × weight 33 × depth 30cm. Size range: 115V, 220–240V.

PRICE CODE: 2

LEHMAN HARDWARE & APPLIANCES INC.
P.O. Box 41
4779 Kidron Road
Kidron
OHIO 44636
USA

NO. 60 CAST IRON POWER MILL

This fast, small grinding mill grinds roots, bones, soybeans, grains etc. The machine itself has a thrust bearing and an extra heavy feed auger. Throughput is 136kg per hour and dimensions are: Height 63.5 × Width 30cm.

PRICE CODE: 1

LEHMAN HARDWARE & APPLIANCES INC.
P.O. Box 41
4779 Kidron Road
Kidron
OHIO 44636
USA

Powered plate mills also manufactured by:

JR ARAUJO IND E COM DE MAQS LTDA
Division Internacional
Rua General Jardim 645
SAO PAULO SP
CEP 01223
BRAZIL

RED STAR MACHINE WORKS OF JIANGXI
Bong Xiang Si
Jiangxi
CHINA

PENAGOS HERMANOS & CIA. LTDA
Apartado Aereo 689
Bucaramanga
COLOMBIA

PRESIDENT MOLLERMASKINER A/S
Springstrup
Box 20
DK-4300 Holbaek
DENMARK

MASKINFABRIKKEN SKIOLD SAEBY
Kjeldgaardsvej
P.O. Box 143
SAEBY
DK 9300
DENMARK

ALVAN BLANCH DEVELOPMENT CO. LTD
Chelworth
Malmesbury
Wiltshire
SN16 9SG
UK

AGRICULTURAL ENGINEERS LTD
Ring Road West Industrial Area
P.O. Box 12127
ACCRA-NORTH
GHANA

C.V. KARYA HIDUP SENTOSA
Jl. Magelang 144
Yogyakarta 55241
INDONESIA

DANDEKAR BROTHERS (ENGINEERS & CO.)
Shivaji Nagar, Factory Area
SANGLI 416 416
INDIA

RAJAN UNIVERSAL EXPORTS(MFRS) P LTD
Raj Buildings 162, Linghi
Chetty Street, P. Bag No. 250
MADRAS 600 001
INDIA

CECOCO
P.O. Box 8
Ibaraki City
OSAKA 567
JAPAN

AGRO MACHINERY LTD
P.O. Box 3281
Bush Rod Island
Monrovia
LIBERIA

MAQ INDUSTRIAL PARA LA PULVERIZACI.
Ruben M Campos 2762
Col Villa de Cortes
03530 MEXICO DF
MEXICO

MAQUINARIA PARA MOLIENDAS Y MEZCLAS
Calle Playa Langosta 25
Col Refora Iztaccihuatl
MEXICO DF CP 08840
MEXICO

ALMEDA COTTAGE INDUSTRY
2326 S. Del Rosario St
Tondo
MANILA
PHILIPPINES

AGROMAC
449 1/1 Darley Rd
Colombo 10
SRI LANKA

USA ECONOMIC DEVELOPMENT CO. LTD
56/7 Thung Song Hong Sub Dist.
Bangkhen
Bangkok 10210
THAILAND

C.S.BELL CO.
170 W Davis Street
Box 291
Tiffin
OHIO 44883
USA

RR MILL INC.
45 West First North
Smithfield
UTAH 84335
USA

RUDARSKO METALURSKI KOMBINAT
Ulica 29
Novembra br. 15
Kostajnica
79224 Bos
YUGOSLAVIA

TONY ELECTRICAL MANUFACTURING
688 Cooleen Road
New Ardbennie
HARARE
ZIMBABWE

41.2 Roller mills

Manual roller mills

TRIPLE ROLL MILLS MODELS 1 AND 2

These mills are used to disperse solids in liquid media. Model 1 can be supplied with either porcelain or stainless steel rolls, while Model 2 has either steel or stainless steel rolls which can be supplied hollow for heating or cooling. Throughputs: (Model 1): 2 litres per hour, (Model 2): 7 litres per hour. Dimensions: depth 48.2 × width 35.6 × height 35.6cm.

PRICE CODE: 4

ADELPHI MANUFACTURING CO. LTD
Olympus House
Mill Green Road
HAYWARDS HEATH
West Sussex
RH16 1XQ
UK

ROLLER MILL, UNATA 8000

This crushes oilseed cells to increase oil yields. Throughput: (Sunflower): 60kg per hour (kernel): 25kg per hour. Dimensions: 159 × 77 × 73 cm.

PRICE CODE: 3

UNION FOR APPROPRIATE TECHNICAL ASSISTANCE
G. Van den Heuvelstraat 131
3140
RAMSEL
BELGIUM

ROLLER MILL KIT, UNATA 8001

This flattens seeds before pressing. It is operated manually by two flywheels and consists of two steel flattening rollers with an adjustable gap. Oil output can be increased by 10–20% by the flattening process. Throughput: 25–60kg per hour.

PRICE CODE: 3

UNION FOR APPROPRIATE TECHNICAL ASSISTANCE
G. Van den Heuvelstraat 131
3140
RAMSEL
BELGIUM

SUGAR-CANE JUICE EXTRACTOR

This manually-operated machine runs at 175rpm. It can also be powered by an electic or gasoline motor. Throughput: 100 litres juice per hour. Dimensions: 85 × 40 × 100cm.

PRICE CODE: 2

CIMAG – COM. E IND DE MAQUINAS AGRICOLA
Rua St Terezinha 1381
13970 ITAPIRA SP
BRAZIL

Roller mills

MINI-TYPE DAL MILL

This mill consists of a pre-cleaning sieve, a sheller/roller, a dal separating sieve, a polisher, a dal grading sieve, an aspirator system, an elevator and an oiling drum machine. Throughput: 340–408kg per hour. Dimensions: length 48 × width 46 × height 51cm. Mills of various power levels available: 7.5, 15, and 22kW (10, 20, 30hp).

PRICE CODE: 4

DAHANU INDUSTRIAL WORKS
PB No. 12 Masoli
Dhanu Road 401 602
DIST THANE (MAHARASTRA)
INDIA

ROLL CRUSHERS

This is used to crush dry products such as biscuits and bread. The rolls are supplied as either smooth or corrugated, and are operated at the same or differential speeds. Depending on the product specification, the rolls are assembled in single, double or triple configurations.

PRICE CODE: 4

GLEN CRESTON LTD
16 Dalston Gardens
STANMORE HA7 1DA
UK

GRANOMILL GRM 40

This machine is suitable for almonds, cocoa nibs, peanuts and hazelnuts. It is designed to grind materials into creamy pastes. Steel heads are agitated in a grind vessel, where impact, attrition or rolling ensures that the required particle size is achieved. Power requirements are a 4hp motor and there is a capacity of 3 litres or 20kg per hour.

PRICE CODE: 4

ALUMCRAFT AGENTINA S.C.A
Mercedes 568
1407 BUENOS AIRES
ARGENTINA

ROOT AND TUBER GRINDER, CNTA 2

This is a grinder for roots and tubers. It is powered by a 3CV engine. Dimensions: 105 × 86 × 86cm.

PRICE CODE: 2

MENEGOTTI – METALURGICA ERWINO
Menegotti Ltda
Rua 590 Erwino Menegotti 381
89250 JARAGUA DO SUL
BRAZIL

VERTICAL SUGAR-CANE CRUSHER

This machine has three crushing rollers, all of which are adjustable externally. It weighs 220kg in total. The power requirement is animal power equal to 1hp. Capacity 4–6 tonnes per day.

PRICE CODE:

HORIZONTAL SUGAR-CANE CRUSHER

This crusher has a 4.5kW (6hp) power requirement and consists of the following parts: high grade semi-steel grooved rollers and touch steel axles; cast iron gears and brass bearings in adjustable cast iron holders. The major rollers have a speed of 15rpm, and the driving pulley operates at 165 rpm. Pulley diameter: 96.5cm. Gross weight: 520kg. Throughput: 417kg cane per hour. Five other models are also available, with powers of up to 12kW (16hp), and throughputs of up to 1.7 tonnes per hour.

PRICE CODE: 3–4

Both the above crushers manufactured by:

PENAGOS HERMANOS & CIA. LTDA
Apartado Aereo 689
Bucaramanga
COLOMBIA

SUGAR-CANE CRUSHER

This machine is designed to squeeze the juice from sugar-cane. The cane is fed between rollers which proceed to crush the cane and squeeze out the juice. The crusher is electrically powered, requiring a 2.2kw/3hp, 3-phase 1500rpm motor. It has a capacity of 20 litres per hour. The dimensions are $128 \times 380 \times 34$cm.

PRICE CODE: 2

TANZANIA ENGINEERING &
MANUFACTURING DESIGN
ORGANIZATION
P.O. Box 6111
Arusha
TANZANIA

SUGAR-CANE CRUSHERS

This machine is a three-roller type with an adjustable upper roller. The power requirement is 6–8hp. The throughput is 3.5–4 tons cane in 8 hours. There is a small size, which can be operated either by hand or by a 2hp motor.

PRICE CODE: 3

DIAS & DIAS
690 Negombo Rd
Mabole
Wattala
SRI LANKA

Sugar-cane crushers also manufactured by:

AGRO MACHINERY LTD
P.O. Box 3281
Bush Rod Island
Monrovia
LIBERIA

NAFEES INDUSTRIES
Samundari Road
Faisalabad
PAKISTAN

P.M. MADURAI MUDALIAR, P.M. & SANG
Madurai Mudalliar Road
P.O. Box 7156
BANGELLERE
INDIA

41.3 Hammer mills

HAMMER MILL, JUNIOR MODEL

This mill is designed for grinding coarse material, particularly gari. It comes equipped with floating steel hammers, changeable perforated screens, feeding bin and a bagging outlet. The machine re-

quires a 2.2kW (3hp) electric motor. Throughput: 30–50kg gari per hour. Dimensions: 50×50cm.

PRICE CODE: 4

INDUSTRIAS MAQUINA D'ANDREA S.A.
Rue General Jardim 645
SAO PAULO 01223
BRAZIL

MAIZE GRINDER

This grinder is fitted with breaker bars which receive most of the impact during grinding. Hammers and screens can be easily changed by raising the hinged top. Weight: 128kg. A 4.5–6kW (6–8hp) motor/engine is required. The 'Standard' model requires 11kW (15hp), and produces 320kg maize per hour. Larger sizes are available.

PRICE CODE: 4

H.C. BELL & SON (PVT) LTD
Mutare
ZIMBABWE

MANIK GRINDING MILL

This mill is designed for use with maize and is fabricated from heavy welded steel. The hammers can be reversed four times for extended life, and the grinding chamber screen can be changed in one minute. There is also an exchangeable liner housing the fan, and a detachable feed table, thus facilitating transport. A 6–9kW (8–12hp) motor/engine is required, which runs at 4000 rpm. Dimensions: Height 63.5cm (with cyclone 168cm). Length complete with feed tray: 96.5cm. Width complete with suction bend: 61cm. Overall weight: 63.5kg. Throughput: 91–181kg per hour.

PRICE CODE: 4

MANIK ENGINEERS
P.O. Box 1274
Arusha
TANZANIA

GRAIN HAMMER MILL

This free-standing hammer mill has three operating speeds. The design maximizes impact velocities. In addition, the mill can be dismantled and assembled quickly and easily for cleaning and transportation. Throughput: 80kg per hour.

PRICE CODE: 3

DEPARTMENT OF AGRICULTURAL
ENGINEERING
Faculty of Engineering
University of Nigeria
NSUKKA
NIGERIA

ELECTRIC MILL M4 228

This small machine grinds coarse, medium and fine materials in seconds and can be used for products such as poppy seeds, sunflower seeds, shelled nuts, spices and coffee. A safety lid is included.

PRICE CODE: 1

RR MILL INC.
45 West First North
Smithfield
UTAH 84335
USA

CUTTER/KNIFE MILL SML

Various agricultural products (particularly suitable for oily products) can be pulverized with this mill. It comes complete with ten interchangeable screens with mesh diameters ranging from 0.25mm to 10mm for particle size control. A model with speed control is available. Throughput: 5–12kg per hour. Overall weight: 93kg.

PRICE CODE: 4

GLEN CRESTON LTD
16 Dalston Gardens
STANMORE HA7 1DA
UK

Powered hammer mills currently available

Country	Manufacturer	Equipment name	Through-put (kg per hour)	Price code	Power (kW)
Belgium	Ateliers Albert & Co. SA	MASTERMILL	320–450	3	4–5.5
Cameroon	Outils pour les communautes	MAMY RAPID HAMMER MILL	250	3	3.7
Colombia	Penagos Hermanos & CIA. LTDA	HAMMER MILL	90–1500	3	3.7–5
Denmark	President Mollermaskiner A/S	HAMMER MILL	250–600	4	3–7.5
	Maskinfabriken Skiold SAEBY	UNIVERSAL HAMMERMILL DM2	80–100	4	5.6–7.5

115

Powered hammer mills currently available (continued)

Country	Manufacturer	Equipment name	Through-put (kg per hour)	Price code	Power (kW)
India	Kaps Engineers	UM HAMMER MILLS HM 10	300	3	
		VM MIKRO PULVERIZERS	450	4	
		UM MASALA MILLS MMO	30–60	3	7.5
	Rajan Universal Exports (MFRS) P Ltd	GRINDING MILL	400–500	3	3.7
Nigeria	Department of Agricultural Engineering	GRAIN HAMMER MILL	80	2	
Peru	Industrias Technicas 'Dolorier'	HAMMER MILL	250–300	4	3.6
	Famia Industrial S.A.	HAMMER MILL	180–800	4	
Philippines	Appropriate Technology Centre	HAMMER MILL FOR COCONUT	5	2	2.2
	Polygon Agro-Industrial Corp.	HAMMER MILL	300	3	2.2
Tanzania	Manik Engineers	MANIK GRINDING MILL	90–180	3	3.7–22
UK	Law-Denis Engineering Ltd	HAMMER MILLS	400–450	4	5.6
	Scotmec (Ayr) Ltd	POPULAR HAMMERMILL	100–200	4	
Zimbabwe	H.C.Bell & Son (Pvt) Ltd	MAIZE GRINDER	320	3	11

41.4 Colloid mills

COLLOID MILL, SMC MJ

This will reduce liquid suspensions to a finer particle size, operating on the principle of high speed, fluid sheer. The material flows through a two-stage grooved conical rotor and a grooved conical stator. The high speed (2800rpm) subjects the fluid to a high shear. Motor power: 5hp (0.37kW). Throughput: 150–1500kg per hour.

PRICE CODE: 3

SREENIVASA ENGINEERING WORKS
2-1-460/1
Nallakunta
HYDERABAD 500 044
INDIA

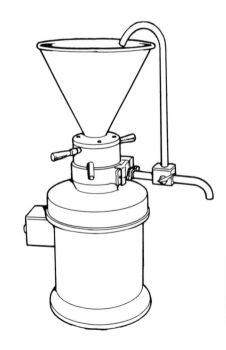

COLLOID MILL

This stainless steel grinder has two speeds: 985rpm or 1970rpm. The 15 litre hopper can be fitted with a jacket for cooling or heating. Throughput: 45–68kg per hour. Details of other models are available on request.

PRICE CODE: 4

ADELPHI MANUFACTURING CO. LTD
Olympus House
Mill Green Road
HAYWARDS HEATH
RH16 1XQ
UK

LABORATORY KEK BETAGRIND

This small-scale production cone mill is particularly suitable for heat sensitive, fatty or sticky substances. Material can either be hand- or continuously-fed into the machine. Dimensions: length 50 × height 53.4 × width 30cm.

PRICE CODE: 4

KEMUTEC GROUP LTD
Hulley Road
Hurdsfield Industrial Estate
MACCLESFIELD
SK10 2ND
UK

MODEL 197A

This machine is for milling sticky, difficult substances. It has a direct drive unit, and requires 0.75–2.2kW (1–3hp). Throughput: 113–122kg per hour. Dimensions: 33 × 84 × 97cm.

PRICE CODE: 4

YTRON QUADRO (UK) LTD
Times House
179 Marlowes
HEMEL HEMPSTEAD
HP1 1BB
UK

COLLOID MILL PUC 60

This mill will produce thin-liquid and paste-like products such as: creams, mayonnaise and fruit juices. The product is fed into the operating area of the stator, which has a speed of 300rpm, by a feed device. The product is processed by high sheer, pressure and friction between the stator and the rotor. Maximum effect depends on: the milling gap, viscosity, feed size and the hardness of the product which is to be processed. The machine has a drive rating of 2.2kW (3hp). All parts are made from stainless steel. Dimensions: 43 × 26 × 68cm.

PRICE CODE: 4

PROBST & CLASS GMBH & CO. KG
Postfach 2053
Industriestrasse 28
RASTATT
7550
GERMANY

BALL MILL

This carries out dry or wet grinding of material to obtain fine particles. The process is carried out in sealed porcelain or steel containers. The roller runs at a speed of 0–45rpm. Dimensions: length 76 × width 42cm. Size range: 0.5 – 30 litres capacity.

PRICE CODE: 3

ADELPHI MANUFACTURING CO. LTD
Olympus House
Mill Green Road
HAYWARDS HEATH
RH16 1XQ
UK

 ## 42.0 Mincers

MEAT MINCERS

This meat mincer is suitable for small scale use. All contact parts are stainless steel, apart from the cutting knives and discs which are made from high carbon steel. There is one cross shaped cutting knife and two cutting discs. The discs have 0.6cm or 0.3cm holes. There are two models available which each require 0.75kW (1hp) 3 phase or 1.1kW (1.5hp). Both machines operate at 1440rpm. Throughputs 60–100kg per hour, or 300–400 kg per hour.

PRICE CODE: 2

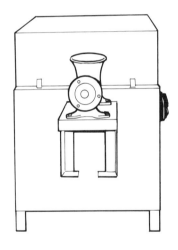

RAYLONS METAL WORKS
Kondivitta Lane
Andheri-Kurla Road
BOMBAY
400 059
INDIA

SEMI-AUTOMATIC MEAT MINCERS

These mincers have most parts made from stainless steel. The power requirements range from 0.25kW (1/3hp) to 1.5kW (2hp). The table below summarizes details of the models available.

Model	Capacity (kg)	Throughput (kg per hour)	Price code
S1	3	25	2
S2	12	60	2
S3	20	120	3

AUTOMATIC MEAT MINCERS

These machines are all floor-standing models. They use a continuous feed system to the cutters by a feed screw. The three sizes available are shown below.

Model	Power kW/hp	Capacity (kg)	Throughput (kg per hour)
S4	3kW/4hp	30	250
S5	5.6kW/7.5hp	70	500
S6	7.5kW/10hp	80	750

PRICE CODE: 4

The above models are manufactured by:

STADLER CORPORATION
P.O. Box 7177
BOMBAY 400 070
INDIA

MEAT MINCER

This stainless steel mincer has a manual piston and uses a three phase electrical motor. Throughput: 720kg per hour.

PRICE CODE: 4

MITCHELL AND JOHNSON PVT LTD
P.O. Box 966
BULAWAYO
ZIMBABWE

MINCER

This minces fresh, tempered or frozen meat. Two models are available which have a continuous forward impulse/reverse switch. Descriptions of the models are as follows:
AE12: powered by a 1.3kW (1.7hp) motor, has four blade knives, two plates and a feed stick.
AF22: powered by a 1.5kW (2hp) motor, and comprising four blade knives, two plates and a cast iron worm.
Throughputs: 230/400kg per hour depending on model.

PRICE CODE: 4

CRYPTO PEERLESS LTD
Bordesley Green Road
BIRMINGHAM
B9 4UA
UK

MEAT MINCERS

Meat can be minced using this equipment when attached to a VARIMIXER. Three sizes are available as follows:
(1) No. 5 – Uses a 20 litre mixer, with a single tool set, plastic tray and stomper. **PRICE CODE: 2**
(2) No. 6 – This uses a double cutting tool, for use with all mixers. **PRICE CODE: 2**
(3) No. 7 – Used with reinforced 40, 60 and 100 litre mixers. This has a double cutting tool. **PRICE CODE: 3**
Throughput: 150, 200, 300 kg per hour.

SPANGENBERG BV
Grootkeuken Professionals
De Limiet 26
4124 PG VIANEN
NETHERLANDS

 ## 43.0 Mixers

43.1 Liquid mixers

LOW TYPE BATCH MIXERS, MODEL 1

This small industrial mixer is available with either a copper or stainless steel inner vessel. An independent motor raises and lowers the agitator. Single, multi or variable speed agitators are available, and insulated jackets can be provided on request. Capacity: 75kg. Dimensions: diameter 60cm, depth 45cm.

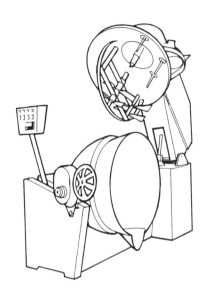

PRICE CODE: 4

BCH EQUIPMENT
Mellor Street
ROCHDALE
OL12 6BA
UK

MANTOO PLANETORY MIXER

The machine can operate at three speeds and has three agitators. In addition to mixing creams and liquids, it is suitable for mixing, whipping, beating and kneading. It uses a planetary action for uniform mixing. Requires 0.75kW (1hp). Capacity: 10kg. Dimensions: 90 × 50 × 120cm.

PRICE CODE: 2

MANGAL ENGINEERING WORKS
Factory Area
PATIALA 147 001
INDIA

LOW MIXING MACHINE

This is designed for use throughout the confectionery trade. There is a lift-out agitator assembly for tilting pans of viscous products. The machine is available with single, double or multiple speed units as requested.

PRICE CODE: 4

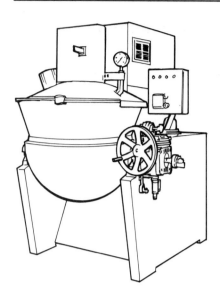

BCH EQUIPMENT
Mellor Street
ROCHDALE
OL12 6BA
UK

TYPHOON PORTABLE AGITATORS

These mixers can be adjusted to any working angle with a single locking device. There are three models in the range, ranging from 19kW (0.25hp) to 2.2kW (3hp).

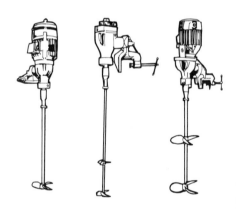

PRICE CODE: 1
PATTERSON INDUSTRIES (CANADA) LTD
250 Danforth Rd
SCARBOROUGH
Ontario M1L 3X4
CANADA

BLENDER

This stainless steel blender has a speed of 3800rpm powered by a 0.25kW (0.3hp) motor. It requires either a 115V 50Hz or 230V 60Hz power supply. Capacity: 10 litres or 12 litres.

PRICE CODE: 2
EQUIPOS HERGO INTEGRALES
Av Division del Norte 2984
C.P 04370 MEXICO DF
MEXICO

K1 BATTER MIXER

A smooth batter can be produced in under a minute using this machine, which comprises a steel bucket and mixer powered by a 0.19kW (0.25hp) motor. Capacity: 10 litres.

PRICE CODE: 3
CRYPTO PEERLESS LTD
Bordesley Green Road
BIRMINGHAM
B9 4UA
UK

DYNAMIC BEATER/BLENDER FT79

This hand blender combines powerful mechanical action with the advantages of a hand tool. The speed can be adjusted using an electronic speed control between 350 and 900rpm to suit the user's technique. Power requirements: 220/240V, 0.25kW (0.34hp). Capacity: 30kg.

PRICE CODE: 1
MITCHELL & COOPER LTD
Framfield Road
UCKFIELD
TN22 5AO2
UK

DYNAMIC LIQUIDISING MIXER PMX83

This hand-held liquidizer is designed for use during cooking. It is housed in stainless steel with a hardened stainless replaceable blade. Power requirements: 220/240V, 0.2kW (0.27hp). Capacity: 45 litres. The range comprises three models: MX79, SMX 450 and PMX83.

PRICE CODE: 2
MITCHELL & COOPER LTD
Framfield Road
UCKFIELD
TN22 5AO2
UK

RASE MINI MIXER

This hand-held mixer is suitable for making soups, purees and sauces. The 0.35kW (0.47hp) motor requires a 220V supply. This size of machine is suitable for 3–8 litres of mixture. MINI, STANDARD and SUPER versions are available.

PRICE CODE: 2
SPANGENBERG BV
Grootkeuken Professionals
De Limiet 26
4124 PG VIANEN
NETHERLANDS

Liquid mixers also manufactured by:

PRACTIMIX S.A.
Oriente 83 No 4311
Col Malinche
MEXICO DF
CP 07880
MEXICO

CRYPTO PEERLESS LTD
Bordesley Green Road
BIRMINGHAM
B9 4UA
UK

ADELPHI MANUFACTURING CO. LTD
Olympus House
Mill Green Road
HAYWARDS HEATH
RH16 1XQ
UK

FABDECON ENGINEERS
138 Damji Shamji Ind. Complex
Off Mahakali Caves Rd, Andheri
BOMBAY 400 093
INDIA

43.2 Solid mixers

VARIMIXERS

This system is used for mixing, kneading whipping etc. The speed system consists of two expanding/contracting pulleys and a number of self-adjusting V-belts. The mixers also have planetary movement for reaching all parts of the bowl. Various sizes are available (12–150 litres). A drive can be fitted for accessory equipment.

PRICE CODE: 4
WODSCHOW & CO. A/S
Industrivinget 6
Postbox 10
2605 Broenby
DENMARK

KENWOOD CHEF

This planetary mixer is suitable for a wide range of operations. It comes complete with a whisk and a dough hook. A range of attachments are available in the Kenwood series. This model has a power requirement of 240 Volts and has a loading of 375 Watts. The capacity of the mixing bowl is 1.35kg flour for yeast dough.

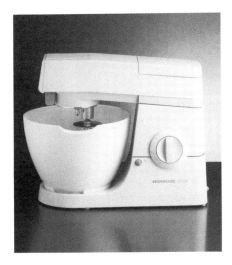

PRICE CODE: 2

The above and many other versions are avilable from:

KENWOOD
Thorn EMI
Domestic Electrical Appliances Limited
Newlane
HAVANT
Hants
UK

HOBART N -50 MIXER

This mixer is suitable for small-scale/kitchen requirements. Using a planetary action, the beater or whip reaches every part of the batch, rotating on its axis opposite to the direction in which it moves around the bowl. It has a single phase voltage. It has a three-speed selective transmission. The bowl has a 5-litre capacity and is fabricated from stainless steel. It can be fitted with a range of different attachments.

HOBART A120 MIXER

This is a bench model with a 12-litre capacity. It has a power requirement of 0.25kw and is available for single and three-phase power supplies. The bowl is fabricated from stainless steel and a full range of beater equipment is available.

PRICE CODE: 2

This and many other models are available from:

THE HOBART MANUFACTURING CO. LTD
Hobart House
51 The Bourne
Southgate
LONDON
UK

43.3 Paste mixers

BEAN PASTE MIXER

This is a steel and brass electric mixer designed for making bean paste. Three models are available with capacities of: 5, 10 and 20kg, and bowl sizes of 51, 102 and 76cm.

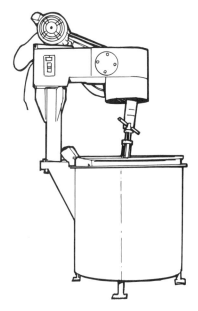

PRICE CODE: 2

KLAUY NAM THAI THOW OP
1505 07 Rama 4 Rd, Wangmai
Patumwan
BANGKOK
10330
THAILAND

43.4 Mixers

SHARIF MIXERS:
DRUM MIXER

This is designed for mixing spices and coating foods with salt, flavours, sugar, etc. It uses the same principles as a tumbler mixer but also uses a hoop which holds the drum at an angle for intensive mixing. The 0.75kW (1hp) motor requires a 415V, 3-phase AC supply. Capacity: 200 litres. Dimensions: 160 × 102.5 × 120cm.

PRICE CODE: 2

FLOUR MIXER

This mixer is made from stainless steel and is designed for mixing different flours. It has a 3.7kW (5hp) motor and a 3 phase motor. Capacity: 2–200 kg .

PRICE CODE: 3

SHARIF ENGINEERING WORKS
Bazar Kharadar
GUJRANWALA
PAKISTAN

 ## 44.0 Moisture analysers

F6 MOISTURE ANALYSER

This measures the moisture content, temperature and weight of products such as coffee beans, cocoa beans, tea, oil seeds, nuts, cereals, rice, pulses, spices and snacks. The unit automatically senses capacitance, weight and temperature of sample. The data is then compared with a commodity moisture curve held in the memory, and the appropriate moisture reading is displayed. The machine is very simple to use. A sample is placed into the measuring cell, the programme is selected and the moisture content is displayed on the LCD display. Power supply: 4 'C' size 1.5 V batteries. Range: 0.5–35% moisture. Dimensions: 32.5 × 16.4 × 12cm.

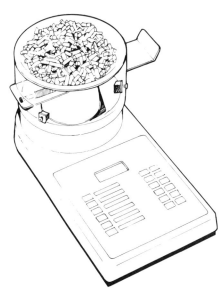

PRICE CODE: 3

SINAR TECHNOLOGY
9 Waterside
Hamm Moor Lane
WEYBRIDGE KT15 2SN
UK

BUTTER MOISTURE BALANCE

This is used for measuring the moisture content of butter.

PRICE CODE: 2

DAIRY UDYOG
C-229A/230A, Ghatkopar Industrial Estate
L.B.S. Marg, Ghatkopar (W)
BOMBAY 400 086
INDIA

INFRA-RED MOISTURE BALANCE

This portable moisture balance is directly calibrated for moisture percentage. Samples of 5–25g can be tested. Accuracy from direct reading = 0.2%.

PRICE CODE: 1

J T JAGTIANI
National House
Tulloch Road
Apollo Bunder
BOMBAY 400 039
INDIA

OSAW DIGITAL MOISTURE METER

The percentage moisture of spices, dehydrated fruit etc. can be measured using this moisture meter which will automatically compensate for ambient temperature. It features a three digit LED display, and has an accuracy of +/– 2% with a moisture range of 8–40%. Requires: 230V AC mains supply.

PRICE CODE: 2

ORIENTAL MACHINERY (1919) PVT LTD
25 R N Mukherjee Road
CALCUTTA 700 001
INDIA

45.0 Moulds and baking tins

45.1 Confectionery and dairy moulds

VOFACHOC HAND MOULDS

These moulds for chocolate are fabricated from impact-resistant hygienic plastic. Various shapes and sizes are available, and they can also be designed to meet individual specifications.

PRICE CODE: 1

G.F.E. BARTLETT & SON LTD
Maylands Avenue
HEMEL HEMPSTEAD HP2 7EN
UK

BUTTER MOULDS

This is a small hand operated block mould for forming butter into rectangular blocks.

PRICE CODE: 1

DAIRY UDYOG
C-229A/230A, Ghatkopar Industrial Estate
L.B.S. Marg, Ghatkopar (W)
BOMBAY 400 086
INDIA

HAND-CARVED WOODEN BUTTER MOULDS

These have a carved interior and a smooth exterior, adding a decorative design to butter. Seven models are available with plain or fancy fluted edges. Alternatively melted butter (or ghee) can be poured into the mould, and when it is hard, it is simply turned out.

PRICE CODE: 1

LEHMAN HARDWARE & APPLIANCES INC.
P.O. Box 41
4779 Kidron Road
Kidron
OHIO 44636
USA

ROSETTE IRONS

Designed for making waffle-like cookies. The irons are dipped into the batter and then cooked in hot oil. As the cookie turns a deep golden brown, the iron is removed from the cookie.
Several models are available.

PRICE CODE: 1

PIZZELLE IRONS

Available in a wide range of designs, these irons are designed for use on gas or electric hotplates.

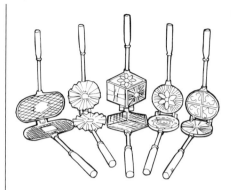

PRICE CODE: 1

Both the above are manufactured by:

VITANTONIO
Dept C 34355 Vokes Drive
Eastlake
OHIO 44094
USA

Confectionery and dairy moulds also manufactured by:

GEBR. RADEMAKER
P.O. Box 81
3640 AB MIJDECHT
NETHERLANDS

45.2 Dough and pastry moulds

ROTARY BISCUIT-MOULDING MACHINE

This forms biscuits into the required shape. Trays are fitted onto a continuous chain. Motors of various powers are available: 0.75kW (1hp), 1.1kW (1.5hp) and 1.5kW (2hp).

PRICE CODE: 3

BIJOY ENGINEERS
Mini Industrial Estate
P.O. Arimpur, Trichur District
KERALA
680 620
INDIA

BREAD MOULDING MACHINE, MDR 2

This machine moulds french and baguette-type bread (from 10g to 2kg). It is made from steel sheets and requires 0.37kW (0.5hp). Throughput: up to 3000 units per hour.

PRICE CODE: 3

SUPREMA EQUIP PARA IND DE PANIFICA
Estrada Municipal SMR 340
N 532/600 Jardim Boa Vista
SUMARE SP
BRAZIL

DOUGH FORMER

This makes up to 74kg of uniformly-sized dough balls per hour.

PRICE CODE: 2

BEIJING No. 1 MACHINE TOOL PLANT
4 Jain Guo Men Wai St
BEIJING
CHINA

HANDAMATIC

This is a hand-operated pie cutting/moulding machine. Optional extras include moulding dies, lidders and platforms for foils. Throughput: 400 pies per hour. Dimensions: 65 × 38 × 37cm.

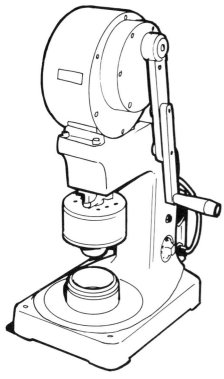

PRICE CODE: 4
CRYPTO PEERLESS LTD
Bordesley Green Road
BIRMINGHAM
B9 4UA
UK

BREAD MOULDING MACHINE

Designed for forming and shaping bread, this machine is suitable for producing dough pieces weighing 400g. Throughput: 500–700 loaves per hour.

PRICE CODE: 3
GARDENERS CORPORATION
6 Doctors Lane
PB No. 299
NEW DELHI 110 001
INDIA

DOUGH MOULDER

This moulds dough into a loaf shape prior to baking. Dimensions: 135 × 70 × 106cm.

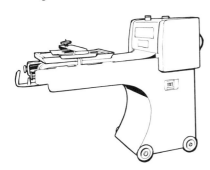

PRICE CODE: 3

KLAUY NAM THAI THOW OP
1505 07 Rama 4 Rd, Wangmai
Patumwan
BANGKOK
10330
THAILAND

RAVIOLI STAMPS

These simple moulds crimp, cut and seal in one motion. They are ideal for pasta and pies.

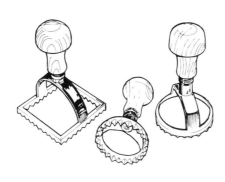

PRICE CODE: 1
VITANTONIO
Dept C
34355 Vokes Drive
Eastlake
OHIO 44094
USA

45.3 Baking tins

MUFFIN PANS

These are used for making decorative biscuits, cakes etc. The following patterns are available: sweetheart, animals, puzzle, teddy bears, harvest fruits, alphabet shapes and a gingerbread house.

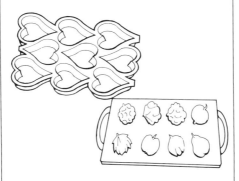

PRICE CODE: 1
LEHMAN HARDWARE & APPLIANCES INC.
P.O.Box 41
4779 Kidron Road
Kidron
OHIO 44636
USA

'EDUCATED' CAKE PAN

This creates different cake shapes using one tin only. The equipment comprises a deep rectangular non-stick cake tin with blocks, which can be placed anywhere inside the cake tin to create a different shape. The cake mixture is then poured around the blocks. Dimensions: 38.5 × 21 × 5cm.

PRICE CODE: 1

MINI BREAD TINS

Are also available from:

LAKELAND PLASTICS
38 Alexandra buildings
WINDERMERE LA23 1BQ
UK

All pans and trays detailed below are available from:

MACKIES PTY LTD
P.O. Box 160
Padstow
NSW 2211
AUSTRALIA

CONFECTIONERY TRAYS

Designed for cakes and buns, these trays include: shell cake trays, American muffin trays, Easter egg/football cake tray, seamless bun tray and mini-loaf pans.

PRICE CODE: 1

CONFECTIONERY PANS

These pans are used for individual cakes and pies. The versions available include: bar loaf pans, fruit cake pans, sponge pans and pie pans.

PRICE CODE: 1

BREAD PANS

The four types of bread pan available are: a three unit bread pan, a single bread pan, corrugated tank loaf pans (three per set), and Vienna pans (three per set).

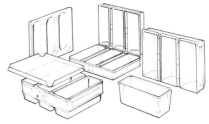

PRICE CODE: 1

Baking trays available from MACKIES

Equipment	Price code	Number of versions available
SPECIALITY ROLL TRAYS	1	3
SPECIALITY BAKING TRAYS	1	5
SPECIALITY LOAF PANS	1	4
VERSATILE STICK/PERFORATED TRAYS	1	2

Other suppliers for MACKIES:

MACKIES PTY LTD
P.O. Box 3
Oundle
PETERBOROUGH PE8 5DS
UK

MACKIES PTY LTD
Unit A, 5 Lorien Place
P.O. Box 82034 Highland Park
AUCKLAND
NEW ZEALAND

 46.0 Ovens

FORCED CONVECTION OVENS

These are designed for bulk cooking, roasting and baking. Two models are available (either gas or electric). The equipment has electronic controls, variable thermostat and a 90-minute timer with manual over ride. Model 320 is heated by burners/elements each side of the compartment, whereas Model 200 is heated on the right hand side of the oven. Capacity: Model 200 = 160 litres, 320 = 275 litres.

PRICE CODE: 4

STOTT BENHAM LTD
P.O. Box 9
High Barn Street
Vernon Works
Royton
OLDHAM OL2 6RP
UK

MODEL FP2E PIZZA OVEN

This compact, electric oven has two cabinets, one on top of the other. Power requirements: 220V, 6kW. Dimensions: 54 × 74 × 68cm. A smaller single-shelved gas pizza oven is also available.

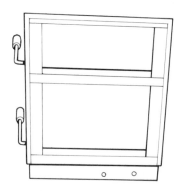

PRICE CODE: 4

CROYDON IND DE MAQS LTDA
Soci Gerente
R 24 de Fevereio 71 F75-77-79
RIO DE JANEIRO, RJ 21040
BRAZIL

BAKERY OVENS

This oven has four chambers and a power consumption of 42kW per hour. Solenoid valves let steam periodically into the oven. Throughput: 200kg per hour. Dimensions: 270 × 320 × 240cm.

PRICE CODE: 4

IND DE FORNOS PERFECTA LTDA
Rua dos Ciclames 460
03164 SAO PAULO SP
BRAZIL

BAKERY/BREAD OVEN

This is designed for the bread baking industry. It has electrical heating and the chambers are well insulated. Throughput: 40kg per hour.

PRICE CODE: 4

SUPREMA EQUIP PARA IND DE PANIFICA
Estrada Municipal SMR 340
N 532/600 Jardim Boa Vista
SUMARE SP
BRAZIL

HAY BOX COOKER

This prototype model comprises a simple box, lined with insulation, into which is placed a pot of previously heated food. The food is cooked in ½–6 hours by the heat retained in the box. There are many advantages for such a simple cooker: for example, the box can save fuel, food can be cooked whilst at work, it is safe near children, it is hygienic, strong, light, portable and the food will never burn.

PRICE CODE: 1

RIM OVEN

This is a prototype small capacity batch oven made from discarded truck rims welded together. A door is welded at the bottom of one, and trays or shelves are welded inside to accommodate bread pans. The oven walls are constructed using clay or mud and fire bricks. Wood is used as fuel.

PRICE CODE: 1

Both the above ovens are available from:

RURAL INDUSTRIES INNOVATION CENTRE
Private Bag 11
KANYE
BOTSWANA

ELECTRIC BAKING OVEN

This oven is insulated with glass wool and heated electrically. Air is circulated in the oven by means of a fan, and the temperature is controlled by a thermostat. The oven uses a 400/440V AC three phase power supply. Size range: Mini, Medium, Large.

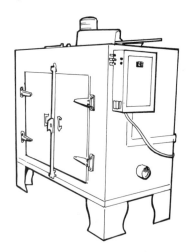

PRICE CODE: 4

BIJOY ENGINEERS
Mini Industrial Estate
P.O. Arimpur, Trichur District
KERALA
680 620
INDIA

BAKING OVEN

Designed for baking bread and cakes, this electrically-heated oven is double-walled and fully insulated. There is also a thermostatic temperature control. Throughput: 300 loaves per hour.

PRICE CODE: 4

GARDENERS CORPORATION
6 Doctors Lane
PB No. 299
NEW DELHI 110 001
INDIA

DRYING OVENS, ELECTRIC & STEAM

These ovens are available as either electric or steam-heated. They are fabricated from mild steel panels and insulated with glass wool. The oven can be heated up to 300°C, and is controlled by a thermostat.

PRICE CODE: 3–4
MASTER MECHANICAL WORKS PVT LTD
Pushpanjali SV Road
Santa Cruz (W)
BOMBAY 400 054
INDIA

HIGH HUMIDITY COOKERS

This cooker uses hot air and steam to heat the product. It is made from stainless steel and is fitted with timer controls and regulators. Cooking is done at a precise temperature, and a smoker can be fitted to the oven. Dimensions: 225 × 125 × 100cm

PRICE CODE: 4
STADLER CORPORATION
P.O. BOX 7177
BOMBAY 400070
INDIA

TWO-DECK GAS BAKING OVEN, GBO 2

Designed for general small-scale batch baking, this oven is heated by a gas burner. Throughput: 50 × 0.5kg loaves per hour. Dimensions: 150 × 95 × 165cm.

PRICE CODE: 4
THAKAR EQUIPMENT CO.
66 Okhla Industrial Estate
NEW DELHI 110 020
INDIA

PIZZA OVEN

This pizza oven has a single tier for direct or pan cooking. It is gas heated and has thermostatic control. Four to six pizzas can be baked per batch.

PRICE CODE: 2
IQBAL ENGINEERING CO.
114 Gulberg Road
LAHORE
PAKISTAN

INDUSTRIAL BAKING OVENS

These small industrial baking ovens will accomodate eight double decker trays of 35 × 46cm per loading. 0.5kg gas per hour is consumed. Dimensions: 87 × 135 × 140 cm.

PRICE CODE: 2
WELL DONE METAL INDUSTRIES INC.
2438 Juan Luna St
Gagalangin
TONDO, Metro Manila
PHILIPPINES

BAKERY OVENS

These stainless steel ovens have one, two or three decks. There is also a spring-loaded hinged door, and heating elements are installed at the top and bottom of the oven. In addition, there is fibreglass insulation at the top and sides of the oven to prevent heat loss, and a fired brick lining at the bottom to aid heat retention. Electric heat control is manual or thermostatic. Requires: 8kW, two-phase, 220V power supply.

PRICE CODE: 4
MITCHELL AND JOHNSON PVT LTD
P.O. Box 966
BULAWAYO
ZIMBABWE

A range of ovens is available from:
OLIVERTONS CATERING EQUIPMENT LTD
Unit 2 Perivale Industrial Park
Horsenden Lane, South Perivale
GREENFORD
UB6 7RX
UK

Ovens are also manufactured by:
GEBR. RADEMAKER
P.O. Box 81
3640 AB MIJDECHT
NETHERLANDS

KLAUY NAM THAI THOW OP
1505 07 Rama 4 Rd, Wangmai
Patumwan
BANGKOK 10330
THAILAND

KGOTETSO OVEN

This is a high-output oven for medium-scale rural production. The oven is constructed from mild steel and it incorporates three compartments: a fire box, baking box and a proving box. It has sliding trays in the baking and proving boxes. A water heater is included in a chimney to utilize the excess heat from smoke emitted from the oven. The baking door is fitted with a temperature gauge. Throughput: 40 loaves per batch.

PRICE CODE: 2
RURAL INDUSTRIES INNOVATION CENTRE
Private Bag 11
KANYE
BOTSWANA

 # 47.0 Packaging equipment

47.1 Sealing equipment

Manual sealing machines

HAND-OPERATED FOIL LID SEALING MACHINE

This hand-operated machine is designed for simple, home use. The operator places the container underneath the heat sealing head, pulls the handle down and, after approximately 1.5 secs, releases the handle to seal the container.

PRICE CODE: 2
ADELPHI MANUFACTURING CO. LTD
Olympus House
Mill Green Road
HAYWARDS HEATH
RH16 1XQ
UK

SEALING MACHINE MK1

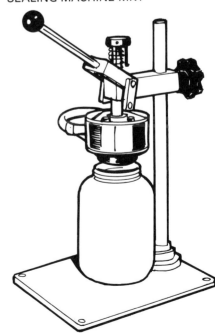

Packaging equipment

This hand-operated heat sealing machine is designed to seal foil lids to containers. It is suitable for small-scale producers, laboratories and R&D establishments. Semi-automatic models are also available. All machines can accommodate a wide selection of containers from yoghurt pots to jerry cans. Throughput: 360–480 containers per hour.

PRICE CODE: 2

CHADWICKS OF BURY LTD
Head Office & Foil Lidding Division
Villiers Street
BURY
BL9 6B2
UK

CAN CLOSING MACHINE

This hand-operated can sealer is flywheel operated, and has an automatic sealing operation, capable of sealing cans up to 10cm in diameter.

PRICE CODE: 2

NARANGS CORPORATION
25/90 Connaught Place
Below Madras Hotel
NEW DELHI 110 001
INDIA

TUBE SEALING CRIMP AND NUMBER MACHINE

This hand-operated machine crimps, seals and numbers batches of plastic tubes. It is suitable for size range 0.95 to 3.2cm. One set of digits is used (0–9) for numbering.

PRICE CODE: 2

J T JAGTIANI
National House, Tulloch Road
Apollo Bunder
BOMBAY 400 039
INDIA

Manual sealing machines also manufactured by:

DAIRY UDYOG
C-229A/230A, Ghatkopar Industrial Estate
L.B.S. Marg, Ghatkopar (W)
BOMBAY 400 086
INDIA

Powered sealing machines

CUP SEALING UNIT

This machine heat seals lids onto plastic cups. Throughput: 800 units per hour. Dimensions: 55 × 40 × 65cm.

PRICE CODE: 2

BRAS HOLANDA
P.O. BOX 1250
CURITIBA PR-80.001
BRAZIL

Cup sealing units also manufactured by:

HAHN & CO
Santo Domingo 588
SANTIAGO
CHILE

FERROSTAL BOLIVIA LTDA
Av Mariscal Santa Cruz
Edif Hansa Piso 12 of 1 Cassila
LA PAZ
BOLIVIA

CHACHOMER S.A. COML. INDL.
Casila De Correo 707
ASUNCION
PARAGUAY

FRAGOL LTDA
Av Libertador Brig Gral
Lavalleja 1641-Esc 203
MONTEVIDEO
URUGUAY

AUTO GRANULAR PACKAGING MACHINE

This machine can pack tea, sugar etc. and can also make bags, seal, cut, count and print. The materials which can be used include: cellophane/polythene, polyester/plated, aluminium/polythene, BOPP film and tealeaf filling paper. 3600 to 4800 bags (of dimensions 7–10 × 5–7.5cm) can be produced per hour. Power requirement: 3kW. Overall dimensions: 60 × 79 × 178cm.

PRICE CODE: 4

CHINA LIGHT CORP FOR FOREIGN
ECONOMIC & TECHNICAL
CO-OPERATION
Tianjin Company
10 Youyi Rd
TIANJIN
CHINA

SEMI-AUTOMATIC FOIL LID SEALING MACHINE

This seals lids onto plastic cups and bottles. It is equipped with pre-set controls which determine temperature, dwell time and pressure. The container and cap are fed into the machine, two buttons are pressed and the pneumatically-driven sealing head descends to seal the container. Up to 600 items can be sealed per hour.

PRICE CODE: 4

ADELPHI MANUFACTURING CO. LTD
Olympus House
Mill Green Road
HAYWARDS HEATH
RH16 1XQ
UK

HM 200 POLYTHENE HEAT SEALER

This sealer is used for sealing packaging. The thermostatically-controlled heater bar has a maximum sealing length of 20.3 cm. The machine is pedal operated for bench mounting and power consumption is 0.25kW. A similar version, the HM 201C is suitable for cellulose films.

PRICE CODE: 1

IMPULSE HEAT SEALER

Two models are available: the HM 1300 with a maximum sealing length of 33cm; and the HM 1800 which has a maximum sealing length of 45.7cm. The machines are bench mounted and are available with single or dual electronic time control.

PRICE CODE: 2

Both the above sealers are manufactured by:

HULME MARTIN LTD
6 Brownlow Mews
Guilford Street
LONDON WC1N 2LD
UK

DAPPER PLASTIC BAG AND HEAT SEALER

This is capable of sealing polythene, polypropylene, coated aluminium foils and cellophane laminates. It can be foot- or hand-operated. Once plugged in, the electrical current (230V) only passes through the element when the jaws of the machine are closed. A roll holder and adjustable bag support are available as optional extras. 25cm and 61cm models are also available.

PRICE CODE: 2

THAMES PACKAGING EQUIPMENT CO.
Senate House
Tyssen House
LONDON
E8 2ND
UK

BAR TYPE IMPULSE HEAT SEALER

This simple machine seals plastic bags by pressing the plastic between heated bars, on which a special tape is placed to prevent the plastic burning. It is designed to use electricity, but alternative sources can be used for heating the bar.

PRICE CODE: 2

THOMAS HUNTER LTD
Omnia Works
Mill Road
RUGBY
CV21 3HT
UK

WRAPID BAR AND L SEALER SYSTEMS

Designed for low volume professional packaging, this machine consists of a film sealer unit to package the product and a hand held heat gun to shrink the film tightly around it. It is suitable for products up to 65cm² with a maximum depth of 10cm. Up to 360 packs can be sealed per hour. Three models are available: Models 14, 16, 26.

PRICE CODE: 3

WRAPID L SEALER

This production sealer offers constant heat. No regular maintenance or replacement parts are required, and a wide variety of products (up to 50 × 60cm, with a maximum depth of 15cm) can be sealed into bags. Up to 1200 packs can be sealed per hour. Models available: Models 15, 25.

PRICE CODE: 3

Both the above sealers are manufactured by:

WRAPS UK
Unit 2, Riverside Park
Station Road
WIMBORNE
BH21 1QU
UK

HAND/PEDAL OPERATED SEALING MACHINE

Designed for small scale production, this has a timer for sealing different thicknesses of film (up to 500 gauge) and different plastic materials. A buzzer indicates when the sealing operation is completed. Throughput: 1000 bags per shift. Size range: 20cm, 30cm and 46cm.

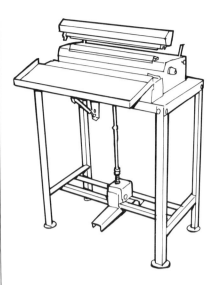

PRICE CODE: 1

Another sealing machine is also available, suitable for most plastics, and capable of producing 3000 bags per shift.

REVO CARRIAGE CONVEYOR MODEL CC-FB

This machine is designed for helping to close jute and paper bags. The operator can start and stop sewing via a foot switch. Weight: 80kg.

PRICE CODE: 1

PORTABLE BAG STITCHER

This portable, electric bag stitcher is finger-operated and produces a single thread stitch. The machine also has a thread cutter. Throughput: 120 bags per hour.

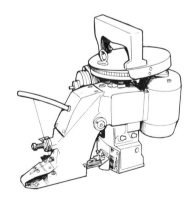

PRICE CODE: 2

A heavy duty bag closer, complete with polypropylene handle, dust proof motor and toothed timing belt, is also available.

The above three items of sealing equipment manufactured by:

SHACO ENTERPRISES
161/163 R R Mohan Roy Road
Pharthana Samaj
BOMBAY 400 004
INDIA

STRIP PACKAGING MACHINE

This machine is ideal for packing chewing gum, tea, coffee etc, and it will use most heat sealable films/foils. First of all, the material is placed into the hopper. Next, a vibratory mechanism draws the material and synchronizes its fall through a chute, channel and release pin into a roller cavity. Heat sealable films from the roll pack seal the material in continuous strips. The machine is powered by a 0.37kW (0.5hp) motor, and seals 24000 to 36000 strips per hour. Size range: S2V, 4V, SP.

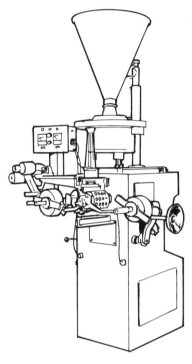

PRICE CODE: 4

SREENIVASA ENGINEERING WORKS
2-1-460/1
Nallakunta
HYDERABAD 500 044
INDIA

TEA PACKER

This tea packing machine can pack 545–655kg per hour.

PRICE CODE: 3

THE COLOMBO COMMERCIAL CO. LTD
121 Sir James Periris Mawatha
P.O. Box 33
COLOMBO 2
SRI LANKA

HAND-OPERATED VAPOUR VACUUM SEALING MACHINE

'Pry off' and 'twist off' caps can be sealed using this machine which comprises a drill stand with lever, frying pan-type sealing head, steam distributor and cock. The container is placed on the base plate, and after steam ejection the cap is placed on the container which retains a vacuum. Pulling down the lever lowers the sealing head to crimp or twist the cap on to the container. Throughput: 360–600 caps per hour.

Packaging

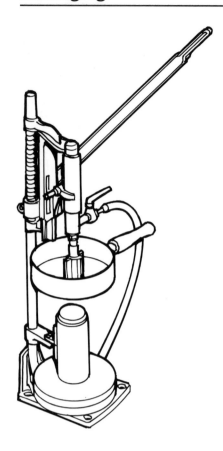

PRICE CODE: 3

FMC CORPORATION
Machinery International Division
P.O. Box 1178
SAN JOSE
California 95108
USA

Powered sealing machines also manufactured by:

AGRICULTURAL ENGINEERS LIMITED
Ring Road West Industrial Area
P.O. Box 12127
ACCRA-NORTH
GHANA

GARDENERS CORPORATION
6 Doctors Lane
PB No. 299
NEW DELHI 110 001
INDIA

CHINAR (PVT) LTD
Plot 33, St 10
Sector 1–9
ISLAMABAD
PAKISTAN

AI PACKING MATERIALS LTD
85–105 Stainsley Road
LONDON E14 6JT
UK

47.2 Capping equipment

OMNIA HAND SEALER

This hand-operated, robust jar sealer has one sealing head which is interchangeable for sizes of between 4.3 and 8.3cm. Jars of up to 12.7cm diameter and 26.7cm height can be sealed (minimum height required is 5.08cm). The unit uses a spring to apply

pressure for the sealing action. This is activated by the handle. Throughput: 1500 jars per hour. Dimensions: height 86 × width 28cm.

PRICE CODE: 2

THOMAS HUNTER LTD
Omnia Works
Mill Road
RUGBY
CV21 3HT
UK

MIRA HEAVY-TYPE CROWN CORKING MACHINE

A hand-operated corking machine.

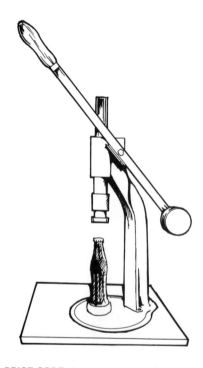

PRICE CODE: 1

ESSENCE & BOTTLE SUPPLY (INDIA) LTD
14 Radha Bazar street
CALCUTTA 700 001
INDIA

CAPPING MACHINE – FLOOR MODEL

This is a hand operated ROPP cap sealer for bottles. The operator places the cap and the bottle on an adjustable base and presses a machine pedal. The machine then knurls the threads of the cap and seams the edge. 600 to 800 bottles can be capped per hour.

PRICE CODE: 1

J T JAGTIANI
National House
Tulloch Road
Apollo Bunder
BOMBAY 400 039
INDIA

AUTOMATIC BUNG PRESSING MACHINE

This machine fits bungs or snap-on caps (both placed manually) on to bottles. The machine automatically orientates the bottle and gently presses down to fit the cap snugly on to the neck of the bottle.

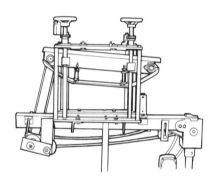

PRICE CODE: 2

MASTER MECHANICAL WORKS PVT LTD
Pushpanjali SV Road
Santa Cruz (W)
BOMBAY 400 054
INDIA

HAND-OPERATED RO/ROPP CAP SEALER

A wide variety of bottles can be closed using this sealer. The machine is suitable for cap sizes of 1.8–7cm, and it can be adjusted to suit bottles ranging from 5 to 30cm in height and 10cm in diameter. Throughput is 600–1200 containers per hour. Hand-operated and semi-automatic versions are available.

PRICE CODE: 2

MASTER MECHANICAL WORKS PVT LTD
Pushpanjali SV Road
Santa Cruz (W)
BOMBAY 400 054
INDIA

CROWN CORKING MACHINE

This hand-operated corking machine consists of: a handle, head, bottle holder, support rod, collar, and foot rest. It can be adjusted according to bottle size. When the handle is pressed, the bottles are corked.

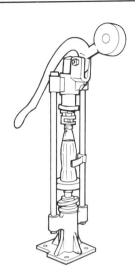

PRICE CODE: 1

RAJAN UNIVERSAL EXPORTS (MFRS) P LTD
Raj Buildings 162, Linghi
Chetty Street, P. Bag No. 250
MADRAS 600 001
INDIA

Powered capping machines

MAUCERI SEMI-AUTO SCREW CAPPER

This bench-mounted screw capping machine can be adjusted to suit the height of containers and can also handle different cap sizes. It comprises a base plate, two support pillars, adjusting screw, head assembly, gear box and motor drive assembly. Operation is by reciprocation of the head driven on continuous cycle from the main drive 0.37kW (0.5hp) motor. Drive to the head is by V-belt and pulley and the machine requires a three-phase, 400/440V electricity supply.

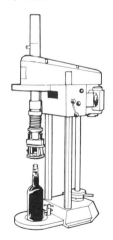

PRICE CODE: 4

Another version, the BENCH PVC CAPSULE HEATING MACHINE T30, is also available, designed to srink PVC capsules onto bottle necks.

VISCOSE CLOSURES LTD
Fleming Way
CRAWLEY
RH10 2NX
UK

RO/ROPP BOTTLE CAP SEALER

This is a semi-automatic cap sealer for small-scale production. The unit is fitted with a 0.37kW (0.5hp) three-phase electric motor. Throughput is 2700–3000 bottles per hour. Hand-operated models are available.

PRICE CODE: 3

RAYLONS METAL WORKS
Kondivitta Lane
Andheri-Kurla Road
BOMBAY 400 059
INDIA

Powered capping equipment also manufactured by:

J T JAGTIANI
National House
Tulloch Road
Apollo Bunder
BOMBAY 400 039
INDIA

MASTER MECHANICAL WORKS PVT LTD
Pushpanjali SV Road
Santa Cruz (W)
BOMBAY 400 054
INDIA

DAIRY UDYOG
C-229A/230A, Ghatkopar Industrial Estate
L.B.S. Marg, Ghatkopar (W)
BOMBAY 400 086
INDIA

RAYLONS METAL WORKS
Kondivitta Lane
Andheri-Kurla Road
BOMBAY 400 059
INDIA

LEHMAN HARDWARE & APPLIANCES INC.
P.O. Box 41
4779 Kidron Road
Kidron
OHIO 44636
USA

NARANGS CORPORATION
25/90 Connaught Place
Below Madras Hotel
NEW DELHI 110 001
INDIA

47.3 Wrapping equipment

GRANULATED CHOCOLATE PACKER

This machine wraps using 007' or 009' aluminium foil. The machine can pack the following sizes: 2.4cm diameter × 1.9cm height, 2.8cm diameter × 1.1cm height or 1.3cm × 1.8cm × 1.35cm. Motor power: 0.55kW (0.74hp). Throughput: 5400–6600g per hour.

PRICE CODE: 4

SHANGHAI DONGGOU BUSINESS MACHINE WORKS
No. 1, Pailouqiao
Pudong
SHANGHAI
CHINA

THE TAVAK L-SEALER

Designed for wrapping breads and cakes, this machine has twin-reel holders to carry both perforated and clear films. Two machine sizes are available: 35.6cm × 61cm or 35.6cm × 83.8cm. The roll holder is moveable on a track allowing the roll position to be adjusted and the tray height within the sealer is adjustable to two positions.

PRICE CODE: 4

TAVAK LTD
Guilford Road
Bisely
WOKING
Surrey
UK

SERIES PACKERS

This machinery is designed for shrink-wrapping. It is automatic (using an auto-feed principle), with low energy consumption and a quick-cooling feature.

PRICE CODE: 4

KAPS ENGINEERS
831 GIDC
Makapura
VADODARA 390 010
INDIA

Wrapping equipment also manufactured by:

WONDERPACK INDUSTRIES PVT LTD
72, 1st Floor, Shivlal Mansion
Lamington Road
BOMBAY 400 008
INDIA

47.4 Vacuum packaging equipment

VACUUM PACKING MACHINE ZBJ350-11

This hand-operated machine removes air from bags before sealing, using two chambers, each with an operating cycle of 20–30 seconds, which work alternately. A vacuum of 1.33 kPa (0.19 psi) is produced.

PRICE CODE: 3

AFFILIATED FACTORY JIANGSU INSTITUTE OF TECHNOLOGY
East Suburb
ZHENJIANG
CHINA

COMPACK VACUUM PACKAGING MACHINES

Products such as cheeses, poultry, smoked meats and ready-to-eat dishes up to 43 × 41cm can be vacuum packed using these machines. The product is placed in the machine and both the evacuation time and sealing time are set on the control panel. A vacuum pump removes the air, and a seal is formed across the bag. An automatic and an electronic version are available.

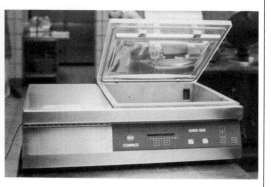

PRICE CODE: 4

KRAEMER UND GREBE GMBH & CO.
Postfach 2149 & 2160
D-3560 Biedenkopf
WALLALL
GERMANY

Vacuum packaging equipment also manufactured by:

NATONG COUNTRY PACKING MACHINE FACTORY
33 Tongyang Naulu
Pingchao Town
NATONG COUNTY
Jiangsu
CHINA

 # 48.0 Pans and kettles

48.1 Small cooking pans

BOILING PAN

This cooks food under pressure of 103 kPa (15psi). It has a vapour valve, steam trap, a filter and a security valve. Capacity: 91 litres.

PRICE CODE: 3

MITCHELL AND JOHNSON PVT LTD
P.O. Box 966
BULAWAYO
ZIMBABWE

STERILIZATION KETTLE MODEL KIT

Oil-yielding fruit (e.g. palm fruit) is placed on top of a perforated base to separate it from the cooking water. A funnel allows control of the water level.

PRICE CODE: 1

UNION FOR APPROPRIATE TECHNICAL ASSISTANCE
G. Van den Heuvelstraat 131
3140 RAMSEL
BELGIUM

WATER BATH

This is used for keeping food at a constant temperature.

PRICE CODE: 1

DAIRY UDYOG
C-229A/230A, Ghatkopar Industrial Estate
L.B.S. Marg, Ghatkopar (W)
BOMBAY 400 086
INDIA

CREAM SETTING PAN

This is a stainless steel pan for small-scale cream production.

PRICE CODE: 1

R J FULLWOOD & BLAND LTD
ELLESMERE
Shropshire
SY12 9DF
UK

NON-STICK PRESERVING PAN

This pan is ideal for making jams. It has two small handles at the sides and one large swinging handle over the top, for easy handling purposes. Capacity: 9 litres. Diameter: 25cm.

PRICE CODE: 1

UNI POT STEAMER

This uses two pans for cooking on one heat source. Stainless steel pans can be placed on top of each other for economical cooking. Capacity: base 2.5 litres, top 1.75 litres. diameter (base) 19.5cm.

PRICE CODE: 1

LAKELAND PLASTICS
38 Alexandra buildings
WINDERMERE
LA23 1BQ
UK

ENERGY-SAVING COOKWARE

This four-in-one-type cooking device consists of four utility pans or skillets, each with a detachable handle, arranged in layers. Once the food is cooked it can stay hot for up to two hours. Dimensions: height 30.5 cm, diameter 25.4 cm.

PRICE CODE: 1

GINHAWA MARKETING INC.
Suite 210 – Regina Building
Escolta
MANILA
PHILIPPINES

48.2 Commercial pans

COATING PAN, 12'

This small pan is used for coating chocolate and confectionery. It has a 0.25kW (0.33hp) motor and a working speed of 30 rpm. Other sizes available: 30, 40.6, 61, 96.5, 106.7, 122, 152.4cm.

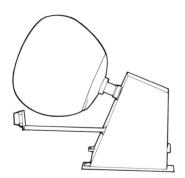

PRICE CODE: 2

OHARA MANUFACTURING LTD
65 Skagway Avenue
TORONTO MIM 3TD
CANADA

TILTING PAN

This bench-mounted tilting pan is electrically heated on single-phase supply. Capacity: 20–50 litres.

PRICE CODE: 4

ADELPHI MANUFACTURING CO. LTD
Olympus House
Mill Green Road
HAYWARDS HEATH RH16 1XQ
UK

RECIPE PAN

This small industrial pan is made from stainless steel and has an outlet tap at the bottom. The lid is attached by a hinge and bracket.

PRICE CODE: 1

BCH EQUIPMENT
Mellor Street
ROCHDALE OL12 6BA
UK

Commercial pans and heaters also manufactured by:

STOTT BENHAM LTD
P.O. Box 9, High Barn Street
Vernon Works
Royton
OLDHAM OL2 6RP
UK

48.3 Commercial steam jacketed pans and kettles

MODELS DJBC 80, 120

These tilting pans are directly steam-heated and jacketed. Capacity: 80, 120 litres.

PRICE CODE: 2

THAKAR EQUIPMENT CO.
66 Okhla Industrial Estate
NEW DELHI 110 020
INDIA

STEAM JACKETED-TYPE TILTING PAN

This tilting pan has an inner pan and outer jacket made from stainless steel and mounted on a mild steel stand. There is also a pressure gauge, safety valve and pressure release valve. Size range: 10, 15, 15, 50kg.

PRICE CODE: 1

NARANGS CORPORATION
25/90 Connaught Place
Below Madras Hotel
NEW DELHI 110 001
INDIA

TILTING PAN

This steam jacketed pan has a manually-operated tilting mechanism and is designed for easy pouring. The pan is fabricated from stainless steel and all pans can be supplied with hinged or lift-off lids if required. A wide range of models are available with capacities ranging from 23–150 litres.

PRICE CODE: 4

BRIERLEY COLLIER & HARTLEY EQUIPMENT LTD
Bridgefield Street
Rochdale
LANCASHIRE
UK

REVOLVING PANS

Jacketed or single-walled coating pans are available. The feed is placed in the stainless steel revolving pan to obtain an even coating of the required substance. Pans in a wide range of capacities are available.

PRICE CODE: 3

In addition, a steam jacketed pan is available, with a tilting mechanism.

BCH EQUIPMENT
Mellor Street
ROCHDALE
OL12 6BA
UK

STEAM JACKETED KETTLE

This kettle is designed for fruit processing. Two models are available: the tilting and the fixed type. In the fixed type the stirring

mechanism is motorized and teflon scrapers prevent the food charring. Unloading is by an outlet at the bottom of the kettle. The inside is made of stainless steel whilst the outer casing is mild steel. Capacity: 250 litres to 500 litres.

PRICE CODE: 4

JYOTHI INDUSTRIES
31 Pampamahakavi Road
BANGALORE 560 004
INDIA

STATIONARY STEAM JACKETED KETTLES

Purees, jams, confectionery etc. can be processed in this range of kettles. The contents of the pans can be drained by stainless steel valves in the kettle base. They are supplied with a pressure gauge safety and wheel valve. Capacity: 23–460 litres.

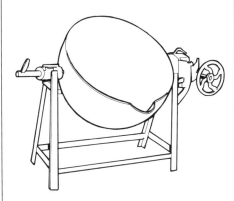

PRICE CODE: 2

RAYLONS METAL WORKS
Kondivitta Lane
Andheri-Kurla Road
BOMBAY 400 059
INDIA

STEAM JACKETED COOKING POT

This pan is designed for making jams, purees, confectionery etc. The pans vary from 6kW (8hp) for the 45 litre model, to 12kW (16.1hp) for the 230 litre model.

PRICE CODE: 3

STAINLESS STEEL INDUSTRIES PVT LTD
P.O. Box ST234
Southerton
HARARE
ZIMBABWE

Steam jacketed pans/kettles also manufactured by:

ADELPHI MANUFACTURING CO. LTD
Olympus House
Mill Green Road
HAYWARDS HEATH
RH16 1XQ
UK

RAYLONS METAL WORKS
Kondivitta Lane
Andheri-Kurla Road
BOMBAY 400 059
INDIA

FMC CORPORATION
Machinery International Division
P.O. Box 1178
SAN JOSE
California 95108
USA

49.0 Pasta machines

Manual pasta machines

RAVIOLI MAKING MACHINE

Ravioli can be made easily using this machine in the following way. Two rolled out sheets of dough, each 8.5cm wide, leave the pasta machine to feed the ravioli maker. A container for holding the stuffing is provided. The stuffing is placed manually onto the dough and covered. The process is continuous until the kneaded mixture in the hopper is exhausted. Throughput: 20kg per hour.

PRICE CODE: 1

LA PARMIGIANA
GB Della Chiesa 10
FIDENZA (Pr)
43036
ITALY

PASTA DRYER

This has over 1.5m² of drying space for freshly-made noodles.

PRICE CODE: 1

LEHMAN also manufacture hand-operated noodle and ravioli makers.

LEHMAN HARDWARE & APPLIANCES INC.
P.O. Box 41
4779 Kidron Road
Kidron
OHIO 44636
USA

ATLAS PASTA MACHINE

This is a hand-operated pasta machine for the household/small scale. The machine has a spaghetti head and stainless steel cutting blades, and can produce noodles up to 0.64cm wide. Recipes are included.
PRICE CODE: 1
RR MILL INC.
45 West First North
Smithfield
UTAH
84335
USA

Pasta machines are also manufactured by:
VITANTONIO
Dept C
34355 Vokes Dr
Eastlake
OHIO 44094
USA

Powered pasta machines

MACARONI MACHINE

Designed for small-scale macaroni production, this machine comprises a mixer, extruder, electrical controls and chilled water cabinet. Throughput: 10–12kg per hour. A larger model is available, throughout 30–50 kg per hour.
PRICE CODE: 3
JINAN AGRICULTURAL MACHINERY
9 Tuwu Road
Shangong
CHINA

PASTA MACHINE TYPE D45 TN

This machine makes spaghetti, short pasta, lasagne etc. It comprises a stainless steel barrel and safety cover and requires 0.75kW (1hp). The diameter of the pasta extruder is 8.9cm. Throughput: 12/16kg per hour.

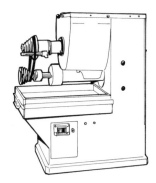

PRICE CODE: 2
LA PARMIGIANA
GB Della Chiesa 10
FIDENZA (Pr)
43036
ITALY

AUTOMATIC RAVIOLI MAKING MACHINE

Different-sized ravioli can be made using this machine by adjusting the amount of stuffing. Power requirement: 0.37kW (0.5hp). Throughput: 20/35kg per hour.
PRICE CODE: 2
LA PARMIGIANA
GB Della Chiesa 10
FIDENZA (Pr)
43036
ITALY

50.0 Pasteurizers

TUBULAR PASTEURIZER

This pasteurizes acid fruit juices. The unit is not manufactured but the university is willing to advise on the setting up of such a pasteurizer. The complete unit comprises: juice tank, pasteurizer with fire box, reservoir, chimney and outlet. Other equipment needed includes: juice extractor, thermometer, bottle capper, bottles and caps. The pasteurizer is topped up by continously checking and pouring water into the reservoir. When it reaches the correct temperature, juice flows from the tank to the pasteurizer, is heated and then pumped through the outlet valve to the filling area.
PRICE CODE: 4
APPROPRIATE TECHNOLOGY DEVELOPMENT INSTITUTE
Private Mail Bag
University of Technology
LAE
PAPUA NEW GUINEA

TUBULAR PASTEURIZER

This prototype equipment has been tested successfully for pasteurizing fruit juice. It consists of a juice tank which feeds juice via a plastic pipe to a stainless steel coil. The coil is immersed in hot oil inside an insulated heater tank. The flow rate of juice is controlled by a valve to achieve the correct pasteurization conditions. It operates continuously at 30–120 litres per hour. Coil size: 15cm diameter made from 2.54cm diameter 18-gauge tube in five sections (bolted together and easily dismantled for cleaning).
PRICE CODE: 1
For drawings and further details contact:
INTERMEDIATE TECHNOLOGY DEVELOPMENT GROUP
Myson House
Railway Terrace
Rugby
Warwickshire CV21 3HT
UK

 51.0 Peeling equipment

Manual peeling machines

FRUIT PEELER

Fruit is positioned on a central pivot. As the handle is turned the fruit is revolved over a stationary knife and the peel is removed.
PRICE CODE: 1
MAQUINARIAS Y AUTOPARTES EL LATINO
Av Grau No 1191
LIMA
PERU

PINEAPPLE SIZER AND CORER

Pineapples can be peeled and cored in the following way. The pineapple is trimmed at both ends and placed into a V-shaped spring-loaded trough as the operator centres the coring tubes. The pineapple is held with the left hand and the manual lever is moved until the coring tube has pierced the fruit. The foot pedal is depressed which raises the stop bar holding the fruit in a coring position. Throughput: 600–720 operations per hour.

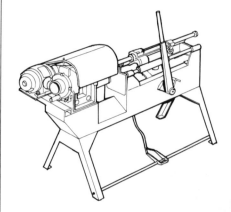

PRICE CODE: 2
FMC CORPORATION
Machinery International Division
P.O. Box 1178
SAN JOSE
California 95108
USA

BEST APPLE PARER (PEELER)

This small, sturdy apple peeler is hand operated. The apple is removed at the end of each cycle. The tool is fabricated from cast iron with a nickel plated apple spike and a replaceable steel blade. The unit will clamp onto surfaces up to 3.2cm thick. Throughput: 360 apples per hour.

PRICE CODE: 1

LEHMAN HARDWARE & APPLIANCES INC.
P.O. Box 41
4779 Kidron Road
Kidron
OHIO 44636
USA

Powered peeling machines

POTATO PEELER

This model has a 0.37kW (0.5hp) electric motor. One cycle takes 3–4 mins. Capacity: 10kg.

PRICE CODE: 3

GARDENERS CORPORATION
6 Doctors Lane
PB No. 299
NEW DELHI 110 001
INDIA

POTATO PEELING AND WASHING MACHINE

This machine has a stainless steel body, cast iron base and a water pipe for washing potatoes. It can be fitted to the table or the floor. Requires: 220/230 50Hz, single-phase electricity supply. Capacity: 100, 200 kg per hour. 0.19kW (0.25hp) and 0.37kW (0.5hp) models are available.

PRICE CODE: 2

IQBAL ENGINEERING CO.
114 Gulberg Road
LAHORE
PAKISTAN

LYE-TYPE PEELER SCALDER

This machine loosens the skins of root crops and some fruits. The unit is made from galvanized material and is equipped with a steam heating coil, quick opening drain, rinsing compartment, and rinse basket. The basket containing the product is placed in the lye peeling section. It is then transferred to the rinsing compartment after a few minutes (the actual time depends on the product being peeled). (See also page 76)

PRICE CODE: 2

FMC CORPORATION
Machinery International Division
P.O. Box 1178
SAN JOSE
California 95108
USA

ELECTRIC POTATO PEELER

This stainless steel, bench top potato peeler has a motor and a V-belt which drives a spindle inside a container. The potatoes are rotated against the abrasive walls. There is a hose entry for cleaning water and an exit for potato peelings.

PRICE CODE: 2

MITCHELL AND JOHNSON PVT LTD
P.O. Box 966
BULAWAYO
ZIMBABWE

ROOT CROP PEELER/WASHER

This machine has a peeling drum with a carborundum surface to peel the food. A water spray continuously removes loosened peel from the machine. Processing is done in batches of 15–20kg. Throughput: 100kg per hour of peeled roots.

PRICE CODE: 2

VISAYAS STATE COLLEGE OF AGRICULTURE
6 Lourdes Street
Pasay City 3129
PHILIPPINES

Powered peeling machines also manufactured by:

THAKAR EQUIPMENT CO.
66 Okhla Industrial Estate
NEW DELHI 110 020
INDIA

CRYPTO PEERLESS LTD
Bordesley Green Road
BIRMINGHAM
B9 4UA
UK

INDUSTRIAS TECHNICAS 'DOLORIER'
Alfredo Mendiola 690
Urbanizacion Ingenieria
LIMA
PERU

 # 52.0 pH meters

This equipment is used to measure the acidity of foods and comes in two basic types: hand-held, electronic meters and larger electric bench-top pH meters. The hand-held type is cheaper and easier to use than the bench models but is less accurate.

BATTERY-OPERATED pH METER 323

This portable pH meter has a range of pH 2–12, with 0.1 pH accuracy.

PRICE CODE: 1

JAGTIANI also manufacture the following pH meters:
• DIGITAL pH METER 353 (mains operated)
• EXPANDED SCALE pH METER (lower scale has 140 divisions)
• pH METER SYSTRONICS TYPE 325 (low cost pH meter measuring pH and mV readings).
• pH METER SYSTRONICS TYPE 324 (supplied with stabilized power supply and a drift-free DC amplifier)

J T JAGTIANI
National House
Tulloch Road
Apollo Bunder
BOMBAY 400 039
INDIA

DELTA 250 RESEARCH pH METER

This measures pH, mV, °C and ion activity. Maximum and minimum limits can be set for any mode, and an alarm will sound if a sample falls outside the required range. Up to five samples can be measured simultaneously via an optional electrode selector.

PRICE CODE: 3

A pocket sized pH meter is also available (the *Ciba Corning* pH meter).

THE NORTHERN MEDIA SUPPLY LTD
Sainsbury Way
HESSLE
HU13 9NX
UK

pH METER DIGITAL

This is a self-contained and portable instrument designed for accurate pH reading. The meter has an LED display for easy reading. Range: pH 0–14. Accuracy: pH +/– 0.01 unit, mV +/– 0.1% of full scale. Temperature compensation: 0–100°C. Power supply: 230V.

PRICE CODE: 2

ESSCO FURNACES (P) LTD
127 Velachery Road
Little Mount, Saidapet
MADRAS 600 015
INDIA

DIGITAL pH METER

This meter measures pH value within the range of pH 0–14 with a resolution of pH 0.01 units. Temperature can be compensated for manually on the LCD display. The unit is powered by a standard 9V battery and comes as a kit containing the instrument, pH electrode and buffers.

PRICE CODE: 2

ELECTRONIC TEMPERATURE INSTRUMENTS
52 Broadwater Street East
WORTHING
BN14 9AW
UK

pH meters are also manufactured by:

WEST METERS LTD
Dept Kom
Western Bank Industrial Estate
WIGTON
CA7 9SJ
UK

THE SCIENTIFIC INSTRUMENT CO. LTD
6 Tej Bahadur Sapru Rd
ALLAHABAD
211 001
INDIA

PHILIP HARRIS SCIENTIFIC
Scientific Supplies Co.
618 Western Avenue
Park Royal
LONDON
W3 0TE
UK

GALLENKAMP EXPRESS
Belton Road West
Loughborough
Leicestershire
LE11 0TR
UK

53.0 Presses

53.1 Fruit and vegetable presses

HYDRAULIC CASSAVA PRESS

This presses grated cassava before or after fermentation. The machine has a high- and low-speed ram operation. There are also two cages so that when one is being loaded, the other is being pressed. During the pressing process, starchy water flows into a container. 0.5 tonnes of cassava can be processed per hour.

PRICE CODE: 2

AGRICULTURAL ENGINEERS LTD
Ring Road West Industrial Area
P.O. Box 12127
ACCRA-NORTH
GHANA

PARALLEL BOARD PRESS

This manual gari press consists of two solid parallel boards. Pulp-filled sacks are placed between the boards and they are then screwed together to create a uniform pressure on the sacks.

PRICE CODE: 1

TECHNOLOGY CONSULTANCY CENTRE
University of Science & Technology
KUMASI
GHANA

FRUIT PRESSES

The fruit is placed into the machine via a hopper. A handle, attached to the machine, is turned to press the fruit and extract the juice. Models available: FPM12, FPM20.

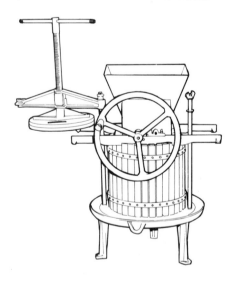

PRICE CODE: 1

HYDRAULIC VINE PRESSES

These manually-operated presses extract juice from soft fruit, e.g. grapes. Hydraulic pressure is used to extract the juice. Various vine presses are available.

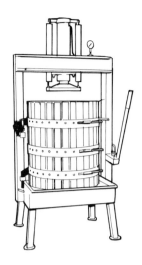

PRICE CODE: 2

Both the above presses manufactured by:

SALINA GEPGYARTO ES
TAMITESTECHNIKA
Vallat
H – 2490 Posztaszblocs
Iskola V.1 PF10
HUNGARY

HAND SCREW BASKET PRESS

This press has a steel frame, wooden tray and basket, and extracts juice from soft fruits.

PRICE CODE: 2

TOMATO AND GRAPE CRUSHER

This machine will crush tomatoes and other soft fruits. The hopper is suitable for hand or pulley drive.

PRICE CODE: 3

Both the above manufactured by:

NARANGS CORPORATION
25/90 Connaught Place
Below Madras Hotel
NEW DELHI 110 001
INDIA

HYDRAULIC BASKET PRESS

This basket press comprises a cylindrical pressing block lined with stainless steel, and has a 51cm diameter. The basket is also made from stainless steel and is perforated. Squeezed juice drains through to a hopper.

PRICE CODE: 2

RAYLONS METAL WORKS
Kondivitta Lane
Andheri-Kurla Road
BOMBAY 400 059
INDIA

UPGRADED TRADITIONAL PRESS

This prototype model, designed to press cassava, is an improvement on the traditional sticks and rope press. It consists of two wooden frames, in between which are placed sacks filled with cassava mash. Threaded rods pass through the bottom frame to fit holes in the top frame, and by tightening the rods the frames can be pressed close together, exerting pressure on the sacks.

PRICE CODE: 1

TIKONKO AG. EXTENSION CENTRE
Methodist Church headquarters
Wesley House
P.O. Box 64
FREETOWN
SIERRA LEONE

GOOD FRUIT CRUSHER

This is a self-feed crusher used for grinding pre–cut apples/small fruit. The crusher is placed across the hopper of the fruit press, the handle is turned and the crushed fruit drops down for pressing. The equipment consists of a cast iron crank, hardwood tub and stainless steel grinding teeth. Dimensions: height 32cm × width 33cm.

PRICE CODE: 1

BEST FRUIT PRESS

This self-contained machine will grind and press all types of fruit. Eight rows of stainless steel teeth are embedded in a hardwood tub. All pulped fruit drops directly into a basket. Basket capacity: 0.035m³.

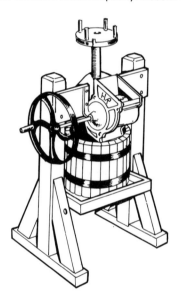

PRICE CODE: 2

Both the above are manufactured by:

LEHMAN HARDWARE & APPLIANCES INC.
P.O. Box 41
4779 Kidron Road
Kidron
OHIO 44636
USA

FRUIT PRESS

This press can be used for grapes and other soft fruit. The press cage is fabricated from a steel frame with hardwood slats. Pressure is exerted by turning a handle at the top of the press. There are four sizes with cage volumes ranging from 8 to 100 litres, with a pressing force from 34.5 to 276kPa (5 to 40psi) and a press stroke from 14.5cm to 30cm. Hydraulic models are also available.

PRICE CODE: 2

RUDARSKO METALURSKI KOMBINAT
Ulica 29
Novembra br. 15
Kostajnica
79224 Bos
YUGOSLAVIA

53.2 Oil presses

UNATA 4202

This press is able to deal with most oil-bearing seeds and nuts. The heated seeds or nuts are placed inside a perforated basket. The oil is extracted by pressure applied manually via the pressing screw. Oil then seeps out onto a collecting tray. Throughput: Peanuts, 9 litres per hour, Palmnut 7 litres per hour.

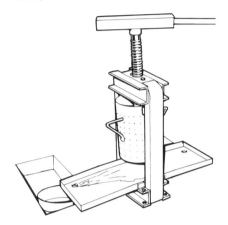

PRICE CODE: 2

In addition, the UNATA 4201 OIL PRESS is also available and is designed for women to use. There are two versions of this model: one for palm fruit and the other for seeds.

UNION FOR APPROPRIATE TECHNICAL ASSISTANCE
G. Van den Heuvelstraat 131
3140
RAMSEL
BELGIUM

MANUAL CALTECH

This manual oil press is designed mainly for palm oil and is hand-operated. The material is fed via a hopper and as the handle is turned the material is pressed against the press head. The oil is collected and the residue is pushed out from the press head. Other versions are available (electrically driven). Extraction rate: 33–37% of available oil.

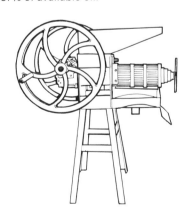

PRICE CODE: 4

OUTILS POUR LES COMMUNAUTES
BP 5946
DOUALA
Akwa
CAMEROON

SCREW PRESS

This manual oil press is designed for the small-scale farmer. It needs two people to operate and it processes 20kg per load.

PRICE CODE: 2

TECHNOLOGY CONSULTANCY CENTRE
University of Science & Technology
KUMASI
GHANA

TINY TECH OIL MILL PLANT

This plant is used for the manufacture of groundnut oil. Groundnuts are fed into a decorticator and the clean kernels are then fed into a cooking kettle which is on top of the expeller. Steam from a baby boiler heats the kernels and some is also sprayed directly onto the kernels so that optimum moisture content is obtained. After this operation, the kernels are fed into the expeller where a pressure exerted by the horizontal screw extracts oil. The residue or cake is emitted at the rear end of the machine, whilst the oil is filtered before leaving. Throughput: 19–31kg per hour.

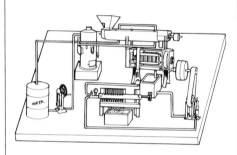

PRICE CODE: 2

GHANI

This is designed for oil production from mustard, surya seeds, etc.

PRICE CODE: 3

BEHERE'S & UNION INDUSTRIAL WORKS
Jeevan Prakash
Masoli
DAHANU ROAD 401 602
Dist Thane
INDIA

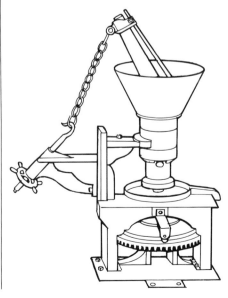

ROTARY OIL MILL

This continuous mill crushes copra, castor, linseed, groundnut etc. It is equipped with a lever and clutch arrangement which enables the removal of oil cakes and reloading without stopping. The oil can be produced within 20–25 minutes. For best performance, the mill should be kept hot by continuous running. Throughput: 35–40 tins each of 16.4kg. Power requirement: 3.7–5.2kW (5–7hp).

PRICE CODE: 2

Both the above mills manufactured by:

RAJAN UNIVERSAL EXPORTS (MFRS) P LTD
Raj Buildings 162, Linghi
Chetty Street, P. Bag No 250
MADRAS 600 001
INDIA

SMALL-SCALE PALM OIL PROCESSING EQUIPMENT

Fresh bunches of oil palm can be processed using this equipment which comprises a stripper, sterilizer, digester, press and a clarifier. Capacity: 0.25 tonnes per hour.

PRICE CODE: 4

NIGERIAN INSTITUTE FOR OIL PALM RESEARCH
Private Mail Bag 1030
BENIN CITY
NIGERIA

KIT PRESS

This hydraulic press is for shea nuts and is suited to the small-scale farmer/co-operative use. There is a fair degree of welding and fabrication skills needed for its manufacture locally. Hydraulic jacks from lorries can be used to press the nuts. Throughput: 5kg of oil per pressing.

PRICE CODE: 4

KIT PRESS

This is a hydraulic press for oilseeds, nuts and fruits. It is suitable for the small-scale farmer. As it generates high pressure, the seeds do not need to be heated, but initial heating will increase the oil yield. Hydraulic jacks from lorries can be used. Throughput: 1–2kg of oil per pressing.

PRICE CODE: 4

Both the above presses from:

ROYAL TROPICAL INSTITUTE
Mauritskade 63
1092 AD
AMSTERDAM
NETHERLANDS

KIT PRESS

This is a manual press for palm oil. The palm fruits are placed inside a perforated basket and the oil is extracted by pressure being applied manually via a pressing screw. Oil then seeps out onto a collecting tray. The skill level required for manufacturing is sophisticated. A larger model is also available. Throughput: 4.5kg of oil per pressing.

PRICE CODE: 3

COCONUT PRESS

This power screw coconut press can press 10 coconuts at a time.

PRICE CODE: 1

Oil presses/Meat presses

Both the above presses from:
ALMEDA COTTAGE INDUSTRY
2326 S. Del Rosario St
Tondo
MANILA
PHILIPPINES

SCREW PRESS

This prototype press is designed for coconut oil extraction. It comprises a hollow vertical metal cylinder with holes punched at its lower end. These allow oil to seep out when ground copra or grated coconut is pressed by a screw mechanism attached to the upper end of the cylinder. This model has a foot-operated lever to aid lifting the cylinder in order to remove pressed material. In addition fruit and cassava starch can be pressed.

PRICE CODE: 2
APPROPRIATE TECHNOLOGY CENTRE
College of Agriculture Complex
Manresa Heights
CAGAYAN DE ORO CITY 9000
PHILIPPINES

BIELENBERG RAM PRESS

This is a hand-operated oil extractor for oilseeds. It is operated by raising and pulling down a long handle which drives a steel ram into a cylinder cage where the seed is pressed. The machine can be operated by one person. When the handle is in an upright position the steel ram is pushed out and the seed is fed from the hopper to the cage. Each time the handle is pressed down, a stream of oil appears from the underside of the cage. 75% of the available oil from sunflower can be extracted.

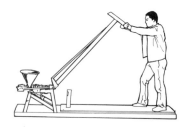

PRICE CODE: 2
ELCT-DAR
Village Sunflower Project
P.O. Box 1409
ARUSHA
TANZANIA

MINI PALM FRUIT DIGESTER (POUNDING MACHINE)

This machine is fitted with self-aligned, sealed and self lubricated bearings. The beaters are made of high carbon steel and the machine can be operated by hand or by motor/engine. Throughput: 100–200kg per hour.

PRICE CODE: 2
AGRICULTURAL ENGINEERS LTD
Ring Road West Industrial Area
P.O. Box 12127
ACCRA – NORTH
GHANA

EPOMIL OIL PROCESSING MACHINE

This machine pounds boiled palm fruits, separating the fibre from the palm nut and extracting the fresh palm oil. It handles a batch capacity of 10kg of boiled palm fruit and takes ten minutes to produce the oil. The machine is incorporated with a gas operated steam cooker and auxiliary screw press. Batch capacity: 10kg fruit.

PRICE CODE: 2
DEPARTMENT OF AGRICULTURAL
ENGINEERING
Faculty of Engineering
University of Nigeria
Nsukka
NIGERIA

Oilseed presses also manufactured by:
KARNATAKA IRON WORKS
Balmatta Road
Near Bendoor Well
MANGALORE 575 002
INDIA

IBG MONTFORTS & REINERS GMBH & CO.
Postfach 200853
D4050
Monchengladbach 2
GERMANY

RAMZAN ENGINEERING WORKS
Maqbool Rd
FAISALABAD
PAKISTAN

KISAN KRISHI YANTRA UDYOG
64 Moti Bhawan
Collector Ganj
Kanpur 208 001
INDIA

AGRICULTURAL ENGINEERS LTD
Ring Road West Industrial Area
P.O. Box 12127
ACCRA-NORTH
GHANA

AGRO MACHINERY LTD
P.O. Box 3281
Bush Rod Island
Monrovia
LIBERIA

AGROMAC LTD
449 1/1 Darley Rd
Colombo 10
SRI LANKA

ALVAN BLANCH DEV. CO. LTD
Chelworth
Malmesbury
Wiltshire
SN16 9SG
UK

53.3 Meat presses

COOKED MEAT PRESS

This press is made from heavy duty cast iron and it has a free-running spindle which is used to press the meat. The dimensions of the ceramic dish are: 16 × 8cm.

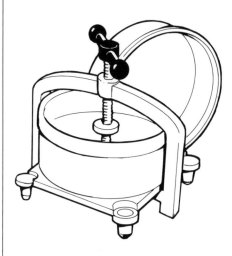

PRICE CODE: 1
LAKELAND PLASTICS
38 Alexandra buildings
WINDERMERE
LA23 1BQ UK

BONZER BURGER MOULD KIT

This is a hand-operated mould, designed to produce 10cm patties approximately 57g in weight which can be quickly pressed between stainless steel plates. The set comprises: mould, 1000 cellulose wraps and a portioner for measuring and handling the meat mixture.

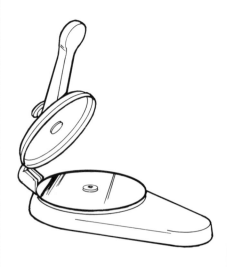

PRICE CODE: 1
MITCHELL & COOPER LTD
Framfield Road
UCKFIELD
TN22 5AO2
UK

54.0 Puffing machines

PUFFED GRAIN MACHINE

This machine will puff grains (eg. rice, corn and sorghum). It can be operated by two people, and is powered by a 4kW (5.4hp) electric motor. Steaming and baking are not required afterwards. Throughput: 18–22kg per hour.

PRICE CODE: 4

YULING MACHINE WORKS
Jianshelu
Guxian County
Henan Province
CHINA

EDIBLE-GRAIN TOASTER

This machine will toast and puff grain. It is powered by a Nissan 1200 motor (petrol engine) and can work for 24 hours with a single operator. Throughput: 1000kg chizitos per day.

PRICE CODE: 4

MAQUINARIAS Y AUTOPARTES EL LATINO
Av Grau No 1191
LIMA
PERU

Puffing machinery also manufactured by:

EL ELEFANTE SR LTELA
Av Aviacian 1525
(Ovalo Av Amola) La Victoria
LIMA
PERU

55.0 Pulpers and juicers

55.1 Steam juicers

MEHU MAIJA

This is a fruit juicer which uses steam to extract the juice. The fruit is placed in a perforated steamer basket. This is stacked on top of a juice kettle and the juice drains through the perforations into a collecting container between the kettle and the steamer basket.

PRICE CODE: 1

LEHMAN HARDWARE & APPLIANCES INC.
P.O. Box 41
4779 Kidron Road
Kidron
OHIO 44636
USA

Steam juicers also manufactured by:

GEBR. RADEMAKER
P.O. Box 81
3640 AB MIJDECHT
NETHERLANDS

RR MILL INC.
45 West First North
Smithfield
UTAH 84335
USA

55.2 Mincing juicers/pulpers

MOULI BABY

This hand-operated machine purees fruit and vegetables. These are placed into a plastic container. A handle turns a knife which churns the fruit, and puree is discharged from the bottom.

PRICE CODE: 1

POTATO RICER

This machine mashes root vegetables, and squeezes the juice out. The vegetable or fruit is placed into the ricer, and the handle is pushed downwards to press the vegetable or fruit. The mash/juice is then extruded through the perforations in the base of the unit.

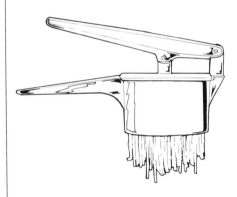

PRICE CODE: 1

Both the above squeezers from:

LAKELAND PLASTICS
38 Alexandra buildings
WINDERMERE LA23 1BQ
UK

JUICE EXTRACTOR

This manually-operated press is for the juice extraction from fruit. The equipment comprises a conical hopper, receptor, precision plate, rod for fixing the hydraulic unit, lever and a protection guard. Two to four operations can be performed per minute. Capacity: 10 litres.

PRICE CODE: 1

DISENOS Y MARQUINARIA JER S.A.
Emiliano Zapata 51
Col Buenavista, 54710
Cuautitlan Itzcalli
MEXICO

VACUUM FRUIT PRESS

This manual fruit press works by exerting pressure inside a specially designed bag. Almost all fruit and vegetables can be pressed with this system. The press kit contains everything that is needed. The fruit is placed inside a special three-layered bag and sealed with a slide grip. Using a simple handpump (which is a reversed bicycle pump), the air can be removed from the system in 30–40 seconds.

The standard press can hold 1.6kg of fruit. A larger capacity press is available and can hold 3.1kg fruit. The system is not intended to be a filter and there will usually be a certain amount of solid material in the juice.

PRICE CODE: 1

CECIL VACUUM SYSTEMS
9 School Place
OXFORD
UK

LEMON REAMER

This hand-operated reamer is used for the pulping of lemons and limes. The pulp and juice are extracted by a pressing action. Length: 15cm.

PRICE CODE: 1

VICTORIO STRAINER

This purees soft fruits and vegetables. No peeling or coring is necessary for this machine, as the juices and fruits are separated from the seeds. The fruit or vegetables are placed in the hopper and the handle is turned. Seeds, skins and cores are continuously separated from the puree. The machine works best with tomatoes and apples but accessories are available for grapes, berries, pumpkins and squash.

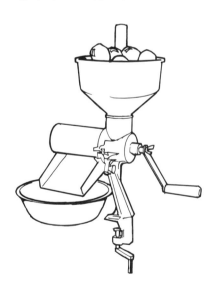

PRICE CODE: 1

APPLE EATER

This is a self-feed apple grinder for high-speed juice extraction. The flywheel is turned and whole apples are self-fed into the grinding mechanism. It has a 1.9cm steel axle mounted on cast iron bearings over 3.8cm long. The mechanism is enclosed in a solid cast iron housing.

PRICE CODE: 1

Pulpers/Juicers

The above are all available from:

LEHMAN HARDWARE & APPLIANCES INC.
P.O. Box 41
4779 Kidron Road
Kidron
OHIO 44636
USA

LIFE STREAM JUICER

This hand-operated juicer extracts the juice from wheat grass, vegetables and fruit. It comprises a stainless steel auger, nylon brushings, a plunger and easy-to-turn handles.

PRICE CODE: 1

RR MILL INC.
45 West First North
Smithfield
UTAH 84335
USA

PULPER BY CITA

This pulper is designed for pulping fruit at the community/village scale. It uses a perforated cylinder which is supported by two plates and held together by four screws. A small diesel or electric 0.75kW (1hp) motor can be used if a 10cm pulley is attached to the pulper axle. The machine can easily be taken apart for washing.

PRICE CODE: 2

MINISTRY OF AGRICULTURE
Dunbars
ANTIGUA

COCOA JUICE EXTRACTOR

This semi-continuous cocoa pulp extractor has a horizontal cylinder with a perforated wall. Rotating blades are mounted on the central axis and these remove the pulp from the seed. Throughput: 99–180kg per hour.

PRICE CODE: 4

EVANDRO FREIRE
Centro de Pesquisas do cacao
Divisao de Technologica Agricola
LX.P7
Itabuna BA
BRAZIL

JUICE EXTRACTOR MODEL EJ4EX

This small industrial juice extractor uses a single phase 0.19kW (0.25hp) motor or a 50/60Hz 110/220V, 550W electricity supply. A sieve is available as an optional extra. Dimensions: length 37cm × diameter 19cm.

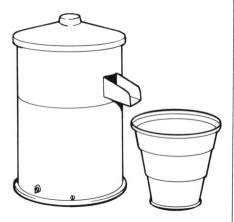

PRICE CODE: 2

LIQUIDIZER LIQ 4

This is used for liquidizing or pureeing fruit and vegetables. The liquidizer has a 0.37kW (0.5hp) motor and requires a 60Hz electricity supply. Included in the equipment are: a cutting knife, a glass bowl with handle and a stainless steel motor housing with speed control. Capacity: 4 litres. Dimensions: 23 × 71 × 23cm.

PRICE CODE: 1

CROYDON IND DE MAQS LTDA
Soci Gerente
R 24 de Fevereio 71 F75-77-79
RIO DE JANEIRO, RJ 21040
BRAZIL

FRUIT AND VEGETABLE PULPER

This pulper is operated by a pulley driven by a 0.37kW (0.5hp) motor. A 35cm sieve is used, giving a throughput of 80kg per hour.

PRICE CODE: 2

Another pulper is powered by a 0.37–1.1kW (0.5–1.5hp) motor, with a throughput of 80–500kg per hour.

COIL TYPE JUICE EXTRACTOR

This extractor has an aluminium juicing head with a stainless steel sieve. The machine is powered by a 0.37kW (0.5hp) motor, processing up to 750 oranges per hour. Alternatively, using a 0.75kW (1hp) motor it can process 2000 oranges per hour.

PRICE CODE: 2
Both the above from:

NARANGS CORPORATION
25/90 Connaught Place
Below Madras Hotel
NEW DELHI 110 001
INDIA

SCREW-TYPE JUICE EXTRACTOR

Designed for medium-scale juice extraction, this machine is driven by a 0.75kW (1hp), three-phase, 440V motor. All contact parts are fabricated from stainless steel and there are two sets of sieves. A hand-operated version is also available. Throughput: 1000 oranges or 800 lemons per hour.

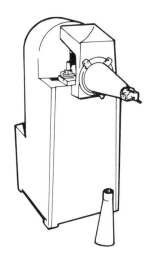

PRICE CODE: 2
RAYLONS METAL WORKS
Kondivitta Lane
Andheri-Kurla Road
BOMBAY 400 059
INDIA

ROTO ROTARY ORANGE JUICER

This is a table-sized automatic orange juicer in a self-contained unit. Oranges are fed into the juice hopper for automatic selection and slicing in half. The orange halves are then mechanically reamed. The seeds are strained and the pulp is compressed to maximize the yield of juice. All waste is deposited in a disposable unit. Throughput: 2640–3960 oranges per hour. Dimensions: length 40.6 × width 22.9 × height 55.9cm.

PRICE CODE: 2
FMC CORPORATION
Machinery International Div
P.O. Box 1178
SAN JOSE
California 95108
USA

NEW WHEATEENA WORKHORSE

Most leafy greens, grasses, herbs and sprouts can be juiced using this machine. The motor is mounted on stainless steel brackets. Throughput: 3.4 litres juice per hour.

PRICE CODE: 2

In addition the following versions are also available:
- WHEATEENA STANDARD ELECTRIC
- POWER UNIT & WHEAT GRASS JUICER

RR MILL INC.
45 West First North
Smithfield
UTAH
84335
USA

Also manufactured by:

BCH EQUIPMENT
Mellor Street
ROCHDALE
OL12 6BA
UK

SPANGENBERG BV
Grootkeuken Professionals
De Limiet 26
4124 PG VIANEN
NETHERLANDS

ROBOT COUPE
10 Rue Charles Delescuze
BP 135
BAGNOLET
93170
FRANCE

Mincing juicers/pulpers also manufactured by:

RAYLONS METAL WORKS
Kondivitta Lane
Andheri-Kurla Road
BOMBAY
400 059
INDIA

NARANGS CORPORATION
25/90 Connaught Place
Below Madras Hotel
NEW DELHI 110 001
INDIA

FOOD PRESERVATION SYSTEMS
1604 Old New Windsor Road
New Windsor
MARYLAND
21776
USA

GARDEN WAY SQUEEZO
1 Mill Street
BURLINGTON
VT 05401
USA

55.3 Coffee pulpers

Manual coffee pulpers

DRUM PULPER

This pulper is designed for small scale coffee pulping. Its principal features include self-lubricating oil-lite brushes. The pulping brush is made from aluminium. In addition to the hand-operated machine, other sizes can be powered by petrol, diesel or electric motors. Capacity: 300kg.

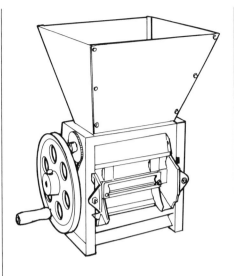

PRICE CODE: 2
DENLAB INTERNATIONAL (UK) LTD
Friary View
40 White Horse Lane
Maldon, Essex
CM9 7QP
UK

IRIMA 67 DISC COFFEE PULPER

This removes the exo- and mesocarp from cherries. The cherries are fed into the hopper with water. The bulbed disc rotates when the handle is turned and, together with the pulping bar, rubs off the outer coatings, leaving a pulp. The pulping bar is straight and has a renewable steel edge. By altering the position of the disc on the shaft, the skin clearance between the disc and the steel edge can be regulated. Throughput: 270–360kg per hour (ripe cherries).

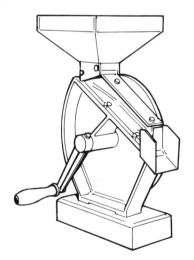

PRICE CODE: 3
JOHN GORDON & CO. (ENGINEERS) LTD
Gordon House
Bower Hill
EPPING
CM16 7AG
UK

HAND-POWERED 'ATOM' PULPER

This pulper is designed for the treatment of Arabica and Robusta coffees. Few adjustments and a minimum amount of water are needed. The machine works on the same

principles as other cylinder pulpers. Throughput: Ripe cherry coffee 54kg per hour.

PRICE CODE: 3
GERICO FRANCE
25 Rue d'Artois
PARIS
75008
FRANCE

Coffee pulpers also manufactured by:

AGRICULTURAL ENGINEERS LTD
Ring Road West Industrial Area
P.O. Box 12127
ACCRA-NORTH
GHANA

KEMAJUAN
Irian Jaya 3
Malang 63118
EAST JAVA
INDONESIA

DENLAB INTERNATIONAL (UK) LTD
Friary View
40 White Horse Lane
Maldon
Essex
CM9 7QP
UK

GERICO UK LTD
Hardy House
Somerset
ASHFORD
TN24 8EW
UK

AGRO MACHINERY LTD
P.O. Box 3281
Bush Rod Island
Monrovia
LIBERIA

Powered coffee pulpers

PULPER AND KNEUZER TYPE S320

This pulper processes red ripe coffee cherry. It comprises: inlet hopper and gate, cylinder with bobbled plate lining (to provide a round surface), pulping knife, outlet chute, rubber knife under pulping knife, pulley and a main shaft. Capacity: 300kg.

PRICE CODE: 3
KEMAJUAN
Irian Jaya 3
Malang 63118
EAST JAVA
INDONESIA

SMALL-SCALE COFFEE PULPER

All types of coffee, except Robusta can be processed using this pulper. A replaceable copper cover is provided and this is protected by a stone catcher in the hopper. The machine requires a power supply of 0.37kW (0.5hp), and two operators. The disc is 33cm in diameter and the pulley is 51cm. Throughput: 150kg ripe cherries per hour.

PRICE CODE: 2
BM ENGINEERING WORKS
PB No. 12
Naidu Street
CHIKMAGALUR 577 101
INDIA

Powered coffee pulpers also manufactured by:

BM ENGINEERING WORKS
PB No. 12
Naidu Street
CHIKMAGALUR 577 101
INDIA

PENAGOS HERMANOS & CIA. LTDA
Apartado Aereo 689
Bucaramanga
COLOMBIA

GERICO UK LTD
Hardy House
Somerset
ASHFORD
TN24 8EW
UK

GERICO FRANCE
25 Rua D'Artois
PARIS 75008
FRANCE

 ## 56.0 Refractometers

7900 REFRACTOMETER

The sugar content of honey can be measured using this hand-held model which operates by placing two to three drops of liquid onto the prism and closing the lid. By looking through the eye-piece, it is possible to read the brix content directly from a gauge. It measures a brix range of 40–85%.

PRICE CODE: 2

SWIENTY
Hortoftvej 16
Ragebol
Sonderborg DK 6400
DENMARK

REFRACTOMETER ABBE IT

This is a general purpose bench refractometer. The sample temperature is measured by a built-in thermistor in the prism and is displayed digitally on a seperate thermometer unit. In addition the following are supplied: integral illuminator, five illuminator lamps (8V 0.15A), refractive index test piece, thermometer and a light source transformer. Power requirements: 240, 50Hz electricity supply. Brix: 0–95%. Dimensions: $13 \times 18 \times 23$cm.

PRICE CODE: 4

REFRACTOMETER SALINITY SI MILL

This hand-held refractometer reads the salt density of various salt solutions. Salinity is measured as salinity per mill units (N°/00). The range of salinity is 0–100°/00. Dimensions: $4 \times 3.5 \times 19.5$cm.

PRICE CODE: 2

REFRACTOMETER N3

This is suitable for products with a high sugar content, such as malt, honey, jam etc. A transparent resin standing up to high temperatures is used on the cover plate. The unit is fitted with a knob which allows the cover plate to be opened more easily when samples of high viscosity are used.

Brix range 58–90%. Dimensions: $3 \times 3.5 \times 13.5$cm.

PRICE CODE: 2

Other versions currently available include the following:
REFRACTOMETER N1 – for the low concentration measurement of sugar contents of fruit, juices, soft drinks and other beverages.
REFRACTOMETER N2 – hand-held and designed for mid-range concentrations.
REFRACTOMETER ATC 1 – hand-held and designed to compensate variable temperatures.
REFRACTOMETER HAND N10 – hand-held and designed for use with products of a low sugar content.

The above all available from:

THE NORTHERN MEDIA SUPPLY LTD
Sainsbury Way
HESSLE
HU13 9NX
UK

SUGAR REFRACTOMETER

This measures sugar content. The following size ranges are available: 0–45%, 40–85%.

PRICE CODE: 1

NARANGS CORPORATION
25/90 Connaught Place
Below Madras Hotel
NEW DEHLI 110 001
INDIA

Refractometers also manufactured by:

AMBRESCO
5600W. Raymond S
INDIANAPOLIS
IN 46241
USA

PHILIP HARRIS SCIENTIFIC
Scientific Supplies Co.
618 Western Avenue
Park Royal
LONDON
W3 0TE
UK

GALLENKAMP EXPRESS
Belton Road West
Loughborough
Leicestershire
LE11 0TR
UK

Honey refractometers:

STEELE & BRODIE
Stevens Drove
Houghton
Stockbridge
Hampshire
UK

 ## 57.0 RH Meters

HUMIDITY METER JENWAY 5060

This is a compact humidity meter using an integral probe. It has the advantage that it can be quickly recalibrated with 75% calibration standard in the form of a screw-on bottle, giving an accuracy of +/- 3% and a resolution of 1%. There is an LCD display and a splashproof touch-sensitive membrane switch. The unit is powered by a 12V

battery. RH range: 15–90%. Dimensions: $12 \times 3.5 \times 2.2$cm.

PRICE CODE: 2

Another model, the HUMIDITY/TEMP METER JENWAY 5500, is also available. This is portable, powered by 5 AA size batteries and measures both wet and dry bulb temperatures to give an RH reading on the LCD display.

THE NORTHERN MEDIA SUPPLY LTD
Sainsbury Way
HESSLE HU13 9NX
UK

RH meters also manufactured by:

GEBR. RADEMAKER
P.O. Box 81
3640 AB MIJDECHT
NETHERLANDS

PHILIP HARRIS SCIENTIFIC
Scientific Supplies Co.
618 Western Avenue
Park Royal
LONDON W3 0TE
UK

GALLENKAMP EXPRESS
Belton Road West
Loughborough
Leicestershire LE11 0TR
UK

58.0 Roasting equipment

ROASTER RA 10015

This will roast coffee, cashew, peanuts and soya. It is equipped with an LPG furnace, pellicle collector, anti-pollution equipment, balcony drawer (to receive the roasted product) and a compartment for the gas cylinder. Capacity: 5kg product per batch.

PRICE CODE: 2

JR ARAUJO IND E COM DE MAQS LTDA
Division Internacional
Rua General Jardim 645
SAO PAULO SP
CEP 01223
BRAZIL

COFFEE ROASTER

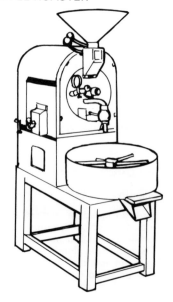

This coffee roaster is suitable for small-scale projects.

PRICE CODE: 1

AGRICULTURAL ENGINEERS LTD
Ring Road West Industrial Area
P.O. Box 12127
ACCRA-NORTH
GHANA

GARI ROASTER

This will roast pulverized and sieved cassava dough. The roasting pan has been constructed with a 3mm iron plate and a chimney is provided to make the roasting process smoke-free.

PRICE CODE: 1

AGRICULTURAL ENGINEERS LIMITED
Ring Road West Industrial Area
P.O. Box 12127
ACCRA-NORTH
GHANA

HAND-OPERATED PEANUT ROASTER

This roaster was developed for peanut roasting on a small scale. The machine consists of a roasting pot, a rotating shaft with an agitator and a stand. 1kg can be roasted in 15 minutes.

PRICE CODE: 1

PAPUA NEW GUINEA UNIVERSITY OF TECHNOLOGY
PM Bag
LAE
PAPUA NEW GUINEA

BRULOIR ROASTER

This will roast cocoa, coffee and cashew nuts. An electric heater is fitted but it is also provided with a space for a kerosene burner. The rotary drum with cross blades gives uniformity and quick discharge of materials into the cooling tray. The automatic stirrer moves and discharges the roasted materials to the final storage tank.

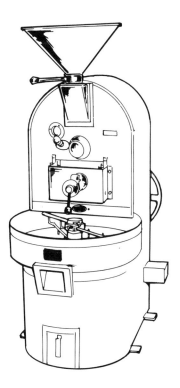

All colour changes can be seen through a viewer. Power requirements: 0.37–1.1kW (0.5–1.5hp) depending on the model. Throughput: 15–100kg per hour. Size range: 4.5, 10, 13, 18, 24, 30kW.

PRICE CODE: 2

RAJAN UNIVERSAL EXPORTS (MFRS) P LTD
Raj Buildings 162, Linghi
Chetty Street, P. Bag No. 250
MADRAS 600 001
INDIA

Roasting equipment also manufactured by:

AGRICULTURAL ENGINEERS LTD
Ring Road West Industrial Area
P.O. Box 12127
ACCRA-NORTH
GHANA

59.0 Rolling equipment

DIAL-A-DEPTH ROLLING PIN

This rolls pastry and pasta, measuring its thickness. The roller is coated with a non-stick finish. The operator chooses the required thickness and then positions the dial at that number. The higher the number, the thicker the pastry. Thicknesses from 0 to 8mm can be measured. Length: 25cm.

PRICE CODE: 1

LAKELAND PLASTICS
38 Alexandra Buildings
WINDERMERE
LA23 1BQ
UK

PAPAD PRESS

This press will press papads from dhal flours. There are two discs: an upper fixed disc and a lower movable disc. The lower disc is connected to a toggle mechanism which is activated by a foot pedal. When pressed, the toggle mechanism moves the lower disc against the upper, causing the papad to be pressed.

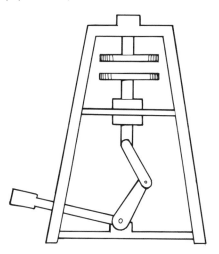

PRICE CODE: 1

CENTRAL FOOD TECHNOLOGICAL RESEARCH INSTITUTE
Food Technology PO
MYSORE 570 013
INDIA

PASTRY ROLLER MACHINE

This rolls pastry using a stainless steel roller on a folding table.

PRICE CODE: 1

MITCHELL AND JOHNSON PVT LTD
P.O. Box 966
BULAWAYO
ZIMBABWE

Rolling machinery also manufactured by:

ASHOK TRADING CO.
18 R N Mukherjee Road
CALCUTTA 700 001
INDIA

JYOTHI INDUSTRIES
31 Pampamahakavi Road
BANGALORE 560 004
INDIA

SHREE RAJALAKSHMI INDUSTRIES
No 61 13th Cross
J.P. Nagar 3rd Phase
BANGALORE 560 002
INDIA

MYSORE PRECISION ENGINEERS
C 123 Industrial Estate
Yadavagiri
MYSORE 570 020
INDIA

60.0 Smoking equipment

MINI SMOKER

This self-contained unit is capable of hot and cold smoking. Air and smoke are drawn horizontally over the products by an integral centrifugal fan driven by a continuously-rated motor. Automatic control of temperature is maintained by a pre-set thermostat, and slide dampers regulate the required smoke density for the product. The air is drawn through a multi-plate diffuser wall, providing an even air flow over the working section. It is then passed by a fan and recirculated by the top duct section. Power requirement: 4.25kW, 240V, 50Hz single phase. Dimensions: length 18.5 × width 6.6 × height 94cm. A MAXI model is also available.

PRICE CODE: 4

AFOS LTD
Springfield Way
Manor Estate
Anlaby
HULL HU10 6RL
UK

CHORKER SMOKER

This large, rectangular smoking oven is both cheap and socially appropriate. The smoking ovens can be constructed in four ways: clay and mud shaped by hand, packed mud faced with cement, clay mud blocks and mortar, or cement blocks with mortar. The top oven is designed to be square, level and flat; the stoke holes should be arched for structural strength; and the oven should be large enough for stoking and removing wood, but not so large as to permit heat and smoke escaping. In addition, the oven should be low, for ease of stacking up to 15 trays, but the fire should be at least 50cm from the lowest tray, hence a 10–20cm fire pit is required for each stoke hole. The smoker is designed so that wooden trays will rest along the midlines of the oven walls. Tray capacity: 15kg fish.

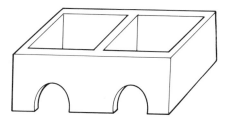

PRICE CODE: 1

NATIONAL COUNCIL ON WOMEN AND DEVELOPMENT
P.O. Box M53
ACCRA
GHANA

DRUM SMOKER

This simple drum smoker can be used by a single person and can be made from scrap materials. It consists of two units: a smoke drum and a fire box. The drum is cleaned and the top is cut out. A fire box is made from a half 210 litre oil drum, and is fitted with a trap door. A smoke pipe with a damper (to control the smoke and fire) is fitted into holes cut in the smoke and fire boxes. Eight round wire mesh trays are required for the smoker. There are two ways of smoking: hot and warm. With hot smoking a short pipe is used between the fire box and the smoke drum. The fish are placed on the trays and are stacked inside the drum to be smoked for three to four hours. Warm smoking uses a long pipe and the damper is only half open. Therefore the fire is not as hot and smoking time takes between four and five hours. Capacity: 15–20kg fish.

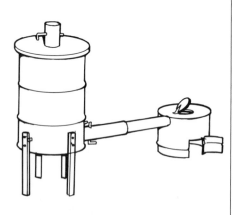

PRICE CODE: 1

CENTRAL FISHERIES RESEARCH INSTITUTE
ZAMBIA

61.0 Sorting equipment

In most very small-scale operations, sorting of raw materials and finished products is done by hand. However, for the larger scale there are a variety of small sorting machines available to grade foods on the basis of size, density or shape. Colour sorting machines are only available for large-scale operations and are expensive.

CONGO COFFEE GRADER

This hand-operated grader separates coffee into three products. It has a cylinder made from detachable perforated sheet with two sizes of perforations, thus giving three products. Spare sheets with different sized perforations can be obtained on request. Throughput: 50kg per hour.

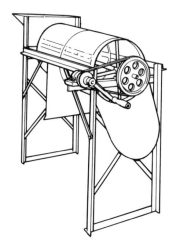

PRICE CODE: 2

The following versions are also available:
UNIVERSAL COFFEE SIZER – provides two or three grades of coffee.
COFFEE TESTING SCREENS – screens are available with seventeen different sizes of round hole perforations.
GRADING MACHINE – separates gari into three grades and can be manually or electrically operated.

JOHN GORDON & CO (ENGINEERS) LTD
Gordon House
Bower Hill
EPPING
CM16 7AG
UK

SFT 500 SIEVING & GRADING MACHINE

This machine is for sieving and grading spices and flours and is designed to provide a gyratory-type vibration. The machine has a 1.1kW (1.5hp) motor which requires a 440V electricity supply. All contact parts are made from stainless steel. Optional extras include, brushing accessories and a stainless steel lid with a charging port. Throughput: 100kg per hour. Dimensions: 100 × 100 × 110cm.

PRICE CODE: 4

IBERIA GRADER MARK II

This single cylinder type grader has screens which are arranged to give a uniform classification of flat beans and perfect separation of peaberries into seven classifications. The grading cylinder is made from special piano wire and perforated steel plates. The feed worm carries the coffee uniformly to the next screens. There is no gearing in the hand powered versions. Each model is available with an electric motor mounted on the machine with a V-motor drive to the cylinder. Four different sizes are available.

PRICE CODE: 4

BRITANNIA MARK II COFFEE GRADER

This double cylinder type grader sorts coffee into nine sizes. Within each grade the flat bean passes over perforated wire screens to ensure that they are of uniform size. They are then graded by their thickness and width. Four different sizes of model are available.

PRICE CODE: 4

The following versions are also available:
ESCOCIA GRADER – separates green coffee beans into five classes.
CASTILLA GRADER – available in three sizes with hand or motive power option, and designed for grading small amounts of coffee.
IBERCALE COFFEE GRADER – similar to the IBERIA GRADER MARK II except that it has outlet spouts and hooks to help bagging.

W M MACKINNON & CO. LTD
Spring Garden Ironworks
Aberdeen
Scotland AB9 1DU
UK
Sorting equipment also manufactured by:

RAYLONS METAL WORKS
Kondivitta Lane
Andheri-Kurla Road
BOMBAY 400 059
INDIA

GARDENERS CORPORATION
6 Doctors Lane
PB No. 299
NEW DELHI 110 001
INDIA

DANDEKAR MACHINE WORKS LTD
Dandekarwadi, Bhiwandi
Dist. Thane
Maharashtra 421 302
INDIA

AGRICULTURAL ENGINEERS LIMITED
Ring Road West Industrial Area
P.O. Box 12127
ACCRA-NORTH
GHANA

STARCH SEPARATOR

This separates and purifies starch produced from maize, cassava or wheat. It is powered by a diesel engine and has a good separation efficiency. Throughput: 15–30kg per hour. Dimensions: 144.7 × 119.6 × 147.6cm.

PRICE CODE: 2

ANQING MARINE DIESEL ENGINE PLANT
38 Du Jiang Lu
Anqing
ANHUI
CHINA

62.0 Storage equipment

PLASTIC HONEY BUCKETS

These are used for storing honey in small to medium quantities. They are white with a handle and have tight fitting lids. The insides are smooth and there are no seams. The bigger versions usually have a gate valve to aid dispensing. Size range: 34, 27, 13kg.

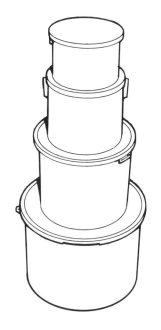

PRICE CODE: 1
NILKAMAL CRATES AND CONTAINERS
5 Rewa Chambers
New Manne Lines
Bombay 400 020
INDIA

CHURNS WITH LIDS

This range of churns includes models which are available with a single handle or handles on both sides. Special pouring churns are also available. Size range: 8, 10, 15, 20, 25, 30, 40, 50 litres.
PRICE CODE: 1
ADELPHI MANUFACTURING CO. LTD
Olympus House
Mill Green Road
HAYWARDS HEATH
RH16 1XQ
UK

STANDARD STORAGE VESSELS

These are stainless steel vessels on wheels in sizes ranging from 20–150 litres.
PRICE CODE: 1–2
ADELPHI MANUFACTURING CO. LTD
Olympus House
Mill Green Road
HAYWARDS HEATH
RH16 1XQ
UK

ALUMINIUM MILK CANS

PRICE CODE: 1

MILK THERMOS CANS

These maintain a constant temperature for storage of milk.
PRICE CODE: 1

MILK CANS WITH 'MUSHROOM' LIDS

PRICE CODE: 1

MILKING PAILS

These are the domed type.
PRICE CODE: 1

MILK AND CREAM DRUMS

These are large pans with handles for milk storage.
PRICE CODE: 1
All the above available from:
DAIRY UDYOG
C-229A/230A, Ghatkopar Industrial Estate
L.B.S. Marg, Ghatkopar (W)
BOMBAY 400 086
INDIA

MILK CANS AND MILK BUCKETS

Various milk cans from 5–20 litres capacity are available for delivery, and 20–50 litres capacity for transport. Milk buckets of capacities 5–15 litres are also available. The transport cans are made from aluminium alloy and are anodyzed, whereas the delivery cans and milk buckets are higher gauge aluminium.
PRICE CODE: 1
DATINI MERCANTILE LTD
Enterprise Road
Box 45483
Nairobi
KENYA

PLASTIC CRATES

This is a range of plastic containers for fruit and vegetables. The containers are made from high-density polyethylene and can withstand temperatures from –30°C to 75°C. The lids and locking clips are available as optional extras. There are eight sizes in the range.
PRICE CODE: 1

SCREEN STORAGE

This is a fly-proof storage unit based on a design from Fiji. The following materials are required for its construction: bamboo or packing cases, wood from the bush, split bamboo, fly-proof plastic or gauze, hinges, screws and nails. A rectangular storage box is made from bamboo, air vents are left in the ends and these are covered with wire gauze. The unit also has two hinged doors at the front which are covered with fly-proof plastic or wire gauze. The legs stand in an ant trap, comprising a small tin holding water or kerosene.

PRICE CODE: 1
ILO – SDSR
P.O. Box 60598
NAIROBI
KENYA

Storage equipment also manufactured by:
LAKELAND PLASTICS
38 Alexandra Buildings
WINDERMERE LA23 1BQ
UK

GEBR. RADEMAKER
P.O. Box 81
3640 AB MIJDECHT
NETHERLANDS

63.0 Temperature measurement equipment

There are three basic types of thermometer: those that contain mercury or alcohol in a glass tube; panels that change colour with different temperatures, and electronic or digital thermometers. The former are cheaper and do not require maintenance but are easily broken which is hazardous in a food processing area. Mercury thermometers should never be used in a food processing room. Electronic thermometers are more expensive and difficult to maintain but are able to measure a wider range of temperatures (eg, from –20°C in a freezer to 400°C in an oven).

Temperature measurement (analogue)

FRIDGE THERMOMETER

This LCD stick-on thermometer measures the temperature of a refrigerator and changes colour accordingly.
PRICE CODE: 1

FREEZER/FRIDGE THERMOMETER

This thermometer hangs, stands or clips into a fridge/freezer. For ease of reading, the temperature zones are colour coded.
PRICE CODE: 1

CANDY/JELLY/DEEP FRY THERMOMETER

This thermometer uses a 15cm probe to measure the temperature of foods whilst cooking. The temperature is displayed on a dial.
PRICE CODE: 1

BRASS SUGAR AND JAM THERMOMETER

This themometer is a traditional brass design. Probe length: 20.5cm.
PRICE CODE: 1

The following thermometers are also available:
(1) MEAT DIAL THERMOMETER
(2) COOKMETER
(3) OVEN DIAL THERMOMETER
(4) WINE & BEER MAKING THERMOMETER
ALL **PRICE CODE : 1**

All the above thermometers from:

WEST METERS LTD
Dept Kom
Western Bank Industrial Estate
WIGTON
CA7 9SJ
UK

THERMOMETER, GENERAL PURPOSE

This thermometer comprises a mercury-filled, toughened bulb, with an anti-roll cap. Seven versions are available, differing in temperature range and graduation. Length: 30.5cm.
PRICE CODE: 1

The THERMOMETER BI-METALLIC DIAL & STEM is also available.
THE NORTHERN MEDIA SUPPLY LTD
Sainsbury Way
HESSLE
HU13 9NX
UK

DAIRY THERMOMETER

This thermometer is designed for butter, cheese and yoghurt production. With the dairy products, accurate temperature control is essential for proper processing. The thermometer floats upright in liquids, is sealed in glass and reads from −10 to 222°F (23–106°C). Length: 21cm.
PRICE CODE: 1

LEHMAN HARDWARE & APPLIANCES INC.
P.O. Box 41
4779 Kidron Road
Kidron
OHIO 44636
USA

DAIRY FLOATING THERMOMETERS

PRICE CODE: 1

DAIRY UDYOG
C-229A/230A, Ghatkopar Industrial Estate
L.B.S. Marg, Ghatkopar (W)
BOMBAY 400 086
INDIA

YOGHURT MAKING THERMOMETER

This measures temperatures between 43°C and 49°C only. Length: 13cm.
PRICE CODE: 1

WEST METERS LTD
Dept Kom
Western Bank Industrial Estate
WIGTON
CA7 9SJ
UK

The following thermometers are available from Lakeland Plastics UK.
(1) MEAT THERMOMETER
(2) PRESERVING THERMOMETER
(3) FOOD PROBE
LAKELAND PLASTICS
38 Alexandra Buildings
WINDERMERE
LA23 1BQ
UK

The following thermometers are available from Electronic Temperature Instruments:
(1) COOKING THERMOMETERS
(2) OVEN THERMOMETERS
(3) FRIDGE THERMOMETERS
(4) PORTABLE CALIBRATOR
(5) GENRAL PURPOSE PROBE 103
(6) HAND-HELD THERMOSENSOR

ELECTRONIC TEMPERATURE INSTRUMENTS
52 Broadwater Street East
WORTHING
BNI4 9AW
UK

Temperature measurement (digital/powered)

POCKET TEST THERMOMETER

This small, portable thermometer has a probe which is placed into the material to be measured. The temperature range is 40–50°C. Length: 15cm.
PRICE CODE: 1

The following thermometers are also available from GBR west:
(1) DIGITAL INDOOR/OUTDOOR THERMOMETER
(2) ELECTRONIC TEST THERMOMETER
(3) ALARM THERMOMETER
WEST METERS LTD
Dept Kom
Western Bank Industrial Estate
WIGTON CA7 9SJ
UK

STICK THERMOMETERS

These thermometers are designed for general purpose temperature measurement. The range includes two probes and one liquid thermometer. The liquid crystal display gives a constant reading, measuring −10°C to 110°C, with an accuracy of 1%.
PRICE CODE: 1

ETI also manufacture the FOOD CHECK THERMOMETER.

ELECTRONIC TEMPERATURE INSTRUMENTS
52 Broadwater Street East
WORTHING
W. Sussex BNI4 9AW
UK

The following thermometers are available from:
(1) THERMOMETER DIGITAL ZEAL
(2) THERMOMETER DIGITEMP
THE NORTHERN MEDIA SUPPLY LTD
Sainsbury Way
HESSLE HU13 9NX
UK

Powered thermometers also manufactured by:

PHILIP HARRIS SCIENTIFIC
Scientific Supplies Co.
618 Western Avenue
Park Royal
LONDON W3 0TE
UK

GALLENKAMP EXPRESS
Belton Road West
Loughborough
Leicestershire LE11 0TR
UK

64.0 Testing, weighing and measuring equipment

64.1 Weighing equipment

ADD AND WEIGH SCALE

This measures the weight of foods. The add-and-weigh facility allows the continuous weighing of ingredients to be carried out in the same bowl. Measurements are carried out in 25g increments. Capacity: 3kg.

PRICE CODE: 1

COMPACT WALL SCALE

The pan is detachable, and can be closed when not in use. Both imperial and metric measurements are shown on the dial. Weights are measured in 20g increments. Capacity: 3kg.

PRICE CODE: 1

Both the above from:

LAKELAND PLASTICS
38 Alexandra Buildings
WINDERMERE
LA23 1BQ
UK

BALANCE OHAUS 1550 SD

This is a two-pan analytical balance for measuring up to 2kg. It comprises a cast alloy base, and features angled beams with centre reading poises, magnetic damping, spring-loaded zero-adjust compensator, and precision-ground knives.

The balance is also equipped with a facility for underbalance weighing. Pan diameter: 15cm.

PRICE CODE: 2

BALANCE SPRING DIAL B14-160

This balance has a hook on which to hang the material to be weighed. The dial is covered in clear plastic and there is a zero adjuster. Weight range: 5–50kg. Length: 26.6cm.

PRICE CODE: 1

Northern Media also supply the following balances:
(1) BALANCE OHAUS 750 SW – a single pan balance, with weighing beams allowing the weighing capacity to be increased.
(2) BALANCE SPRING – capable of measuring weights up to 100kg.
(3) BALANCE SPRING TUBULAR FORM – can weigh up to 1kg with low friction movement.

THE NORTHERN MEDIA SUPPLY LTD
Sainsbury Way
HESSLE
HU13 9NX
UK

BEAM SCALE BD1

These heavy duty bakery scales are designed for everyday baking. They are hand machined, and chrome plated, complete with a sliding poise for accuracy. Additional extras include weights and a stainless steel scoop. Various versions are available, e.g. Scale: 1lb × 1/4 oz (0.45kg × 7g).

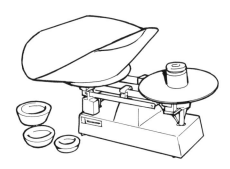

PRICE CODE: 2

HANGING SCALES S30

These heavy duty, hanging scales have an adjustable tare. Capacity: 30kg × 100g.

PRICE CODE: 1

Heavy duty dual dial scales are also available.

DIETETIC SCALES P200

These compact scales are of use wherever precise controlled food or ingredient portioning is required. Capacity: 200 × 2g.

PRICE CODE: 1

The above three scales all from:

MITCHELL & COOPER LTD
Framfield Road
UCKFIELD TN22 5AO2
UK

DAIRY MILK WEIGHING SCALES

These scales are used for weighing milk, and come complete with a tripod and bucket.

PRICE CODE: 1

DAIRY UDYOG
C-229A/230A, Ghatkopar Industrial Estate
L.B.S. Marg, Ghatkopar (W)
BOMBAY 400 086
INDIA

PAN SCALES

Two models are available: one is a simple pan scale with a capacity of up to 4.5kg or 12.5kg (two sizes available). The other is a combined platform/pan scale complete with castors for easy movement. It is capable of weighing up to 1000kg.

PRICE CODE: 3

BALANZAS CONDOR – FR. YEP Y CIA S.A.
Av. Mexico 1535
La Victoria
Lima 13
PERU

2108 COUNTER SCALE

These scales comprise the following: cast iron base, shallow weighing pans, lifting handles and a range of weights. Capacity: 10kg, 20kg.

PRICE CODE: 2

3961 AAG PORTABLE PLATFORM SCALE

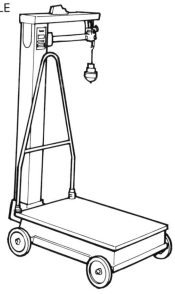

These scales weigh bulky/large materials such as sacks of flour. The equipment includes a cast iron box, steel knife edges, and a platform-on-wheels. Weight of scales: 250kg. Dimensions: 40.6 × 63.7cm.

PRICE CODE: 3

Both the above scales from:

AVERY ZIMBABWE PVT LTD
P.O. Box 392
HARARE
ZIMBABWE

TEFAL ELECTRONIC SCALE

These scales weigh in 1g increments. Any type of bowl or solid food can be placed onto the scales directly. There is also an add-and-weigh feature, and an automatic cut-off. Capacity: 4kg. Dimensions: 15 × 16 × 3cm.

PRICE CODE: 1

LAKELAND PLASTICS
38 Alexandra buildings
WINDERMERE LA23 1BQ
UK

Digital/powered weighing equipment also manufactured by:

FAMIA INDUSTRIAL S.A.
Av. Heroes de la Brena 2790
Ate
Lima 3
PERU

64.2 Measuring cylinders/ jugs

MIXER JUGS

This jug can be used as either a measuring jug or a mixing bowl. It is fabricated from heat-resistant polypropylene and can withstand boiling water. The jug is graduated in pints, fluid ounces and millilitres with an overall capacity of 2 litres. 0.57 litres and 1.1 litres (1 or 2 pints) capacity jugs are also available.

PRICE CODE: 1

LAKELAND PLASTICS
38 Alexandra Buildings
WINDERMERE LA23 1BQ
UK

PYREX MEASURING CYLINDERS

This measuring cylinder has a capacity of 10–250ml.

PRICE CODE: 1

THE NORTHERN MEDIA SUPPLY LTD
Sainsbury Way
HESSLE HU13 9NX
UK

64.3 Lactometers

LACTOMETERS

These lactometers are designed for the testing of milk. Two types are available: Desco (15cm length) which can measure at temperatures up to 16°C; and Quenne (18cm length) which can measure at temperatures up to 60°C in 0.5°C increments.

PRICE CODE: 1

LAKSHMI MILKTESTING MACHINERY CO.
A90 Group Industrial Area
Wazirupa
NEW DELHI
110 052
INDIA

LACTOMETER

This measures the density of milk. Length: 15cm.

PRICE CODE: 1

DAIRY UDYOG
C-229A/230A, Ghatkopar Industrial Estate
L.B.S. Marg, Ghatkopar (W)
BOMBAY 400 086
INDIA

DAIRY UDYOG also manufacture the following equipment:
JARS FOR LACTOMETERS
LACTOMETERS – MERCURY FILLED

64.4 Hydrometers

HYDROMETER

This glass hydrometer measures the specific gravity of liquids. It is calibrated at 16°C and has an accuracy of: +/– 0.002 specific gravity. Length: 19cm.

PRICE CODE: 1

THE NORTHERN MEDIA SUPPLY LTD
Sainsbury Way
HESSLE
HU13 9NX
UK

HYDROMETER

This all purpose glass hydrometer has the following specifications: specific gravity: 0.990–1.170. Brix: 0–38% sugar by weight. Alcohol: 0–22% by volume.

PRICE CODE: 1

LEHMAN HARDWARE & APPLIANCES INC.
P.O. Box 41
4779 Kidron Road
Kidron
OHIO 44636
USA

DIGITAL HYDROMETER 6000

This portable, hand-held hydrometer measures relative humidity levels in food storage areas and in equipment such as dough provers. The display indicates the parameter being measured.

A range of hydrometers is available: Humidity = 5% to 95%. Temperature = 10°C to 70°C.

PRICE CODE: 1

ELECTRONIC TEMPERATURE
INSTRUMENTS
52 Broadwater Street East
WORTHING
BNI4 9AW
UK

64.5 Butyrometers

BUTYROMETER

This measuring container is used to hold the sample when measuring the fat content of milk. A sulphuric acid reagent and a centrifuge are also required. Range: 0–10%, 0–80% fat.

PRICE CODE: 1

LAKSHMI MILKTESTING MACHINERY CO
A90 Group Industrial Area
Wazirupa
NEW DELHI
110 052
INDIA

The following pieces of equipment are each used with butyrometers:

BUTYROMETER STAND

This is for holding butyrometers and is the shaking type.

PRICE CODE: 1

ALUMINIUM WATER BATH

This water bath keeps butyrometers at the correct temperature.

PRICE CODE: 1

DOUBLE CONICAL RUBBER STOPPERS

These stoppers are designed for the temporary sealing of butyrometers.

PRICE CODE: 1

DAIRY UDYOG
C-229A/230A, Ghatkopar Industrial Estate
L.B.S. Marg, Ghatkopar (W)
BOMBAY 400 086
INDIA

64.6 Testing equipment

METHYLENE BLUE APPARATUS

This is used for testing the amount of live bacteria in milk.

PRICE CODE: 1

DAIRY UDYOG
C-229A/230A, Ghatkopar Industrial Estate
L.B.S. Marg, Ghatkopar (W)
BOMBAY 400 086
INDIA

SAMPLERS FOR MILK

These are ladles which are dipped into milk to obtain a sample.

PRICE CODE: 1

DAIRY UDYOG
C-229A/230A, Ghatkopar Industrial Estate
L.B.S. Marg, Ghatkopar (W)
BOMBAY 400 086
INDIA

PIPETTES FOR MILK, ACID AND ALCOHOL

These are used for obtaining small samples of these liquids. Automatic pipettes are also available.

PRICE CODE: 1

DAIRY UDYOG
C-229A/230A, Ghatkopar Industrial Estate
L.B.S. Marg, Ghatkopar (W)
BOMBAY 400 086
INDIA

BRINE METER

This meter can be used to measure the percentage of salt in a solution.

PRICE CODE: 1

GEBR. RADEMAKER
P.O. Box 81
3640 AB MIJDECHT
NETHERLANDS

Brine meters and other testing equipment also supplied by:

PHILIP HARRIS SCIENTIFIC
Scientific Supplies Co.
618 Western Avenue
Park Royal
LONDON
W3 0TE
UK

GALLENKAMP EXPRESS
Belton Road West
LOUGHBOROUGH
Leicestershire
LE11 0TR
UK

65.0 Threshers

Manual threshers

SINGLE OPERATION PADDY THRESHER

Designed for threshing rice paddy, this manually-operated thresher uses the wheel, sprockets and chain of a bicycle for drive. Throughput: 125kg per hour. Dimensions: $100 \times 76 \times 66$cm.

PRICE CODE: 1

**TANZANIA ENGINEERING &
MANUFACTURING DESIGN
ORGANIZATION**
P.O. Box 6111
Arusha
TANZANIA

OLPAD THRESHER (8 DISC)

This machine has serrated discs of 45cm diameter. A comfortable seat is provided for the operator with a back- and foot-rest while the back and front safety guards eliminate injury risk. The harvest is spread on the threshing floor and the machine is drawn round and round thus separating the grain. An extra racking attachment can be fitted for stirring straw during the threshing operation. Models are available with 11, 14 and 20 discs. Weight 92kg. Throughput: 44 to 106kg per hour.

PRICE CODE: 2

COSSUL & CO PVT LTD
123/367 Industrial Area
Fazalgunj
Kanpur
U.P.
INDIA

ROTARY PADDY THRESHER

This is a pedal-operated threshing machine. Paddy sheaves are held in suitable bunches close to the teeth of the rotation cylinder whilst the cylinder is kept in motion by pedalling. The grains are then separated or combed from the ears by wire teeth provided on the spokes of the cylinder. Throughput: 100–200kg per hour.

PRICE CODE: 1

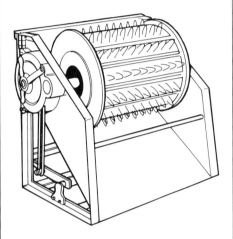

**RAJAN UNIVERSAL EXPORTS(MFRS) P
LTD**
**Raj Buildings 162, Linghi
Chetty Street, P Bag No 250**
MADRAS 600 001
INDIA

Manual threshers also manufactured by:
AGRICULTURAL ENGINEERS LIMITED
Ring Road West Industrial Area
P.O. Box 12127
ACCRA-NORTH
GHANA

CECOCO
P.O. Box 8
Ibaraki City
Osaka 567
JAPAN

AGRO MACHINERY LTD
P.O. Box 3281
Bush Rod Island
Monrovia
LIBERIA

Powered threshers currently available

Commodity	Country	Manufacturer	Equipment name	Through-put (kg per hour)	Price code	Power (kW)
Cereals	Philippines	P.I. Farm Products Inc.	PALAY MINI THRESHER	200	2	10.4–12
	Nigeria	Institute for Agricultural Research	GROUNDNUT THRESHER	120	2	4
	India	Dandekar Brothers & Co.	MULTI-CROP THRESHER	300–400	3	
	Netherlands	Votex Tropical	HIGH SPEED THRESHER	300–2000	4	7
Rice	India	Kisan Krishi Yantra Udyog	PADDY THRESHER	300–500	3	2
	Indonesia	P.T. Rutan Machinery Trading Co.	PORTABLE THRESHER	400–600	3	
	Philippines	JCCE Industries	AXIAL FLOW THRESHER	800–1000	4	7.5
Sorghum	Botswana	Rural Industries Innovation Centre	SORGHUM THRESHER	210	3	
	Costa Rica	Universidad de Costa Rica	THRESHER FOR BEANS	50	2	3.7
Wheat	India	Madho Mechanical Works	MADHO WHEAT THRESHER	300	3	3.7
		Standard Agricultural Eng. Co.	AUTOMATIC FEED WHEAT THRESHER	800–1000	4	15
	Pakistan	United Steel Mills Ltd	WHEAT THRESHER	740	3	11

66.0 Water treatment equipment

WATER TEST KIT LOVIBOND AF 357

This kit is designed for the analysis of free and combined chlorine, pH and fluoride. The test methods are colorimetric using the Lovibond 2000 comparator and accessories. The equipment comes supplied with all the necessary reagents, glassware and instructions in a plastic case.

PRICE CODE: 2

THE NORTHERN MEDIA SUPPLY LTD
Sainsbury Way
HESSLE HU13 9NX
UK

PORTABLE WATER SOFTENER

This portable two-bed water dispenser consists of two FRP/PVC cylinders filled with cation and anion exchange resins. It is mounted on an electrostatically panelled frame which has a valve block. The treated water has dissolved solids below 5ppm, pH 7.5–9 and CO_2 levels below 5ppm.

PRICE CODE: 3

WATRION WATER & FILTER
ENGINEERING P LTD
34A Rameshwar
4 Bungalow Road, Andheri (W)
BOMBAY 400 058
INDIA

UV WATER DISINFECTION UNIT

This is used to treat water by destroying bacteria using ultra-violet light. It uses a single phase, 220V 50Hz power supply, or alternatively it can be battery-operated. Water flow is at 750 litres per hour.

PRICE CODE: 1

WATER ASSOCIATES & EQUIPMENT
P.O. Box HG 423
Highlands
HARARE
ZIMBABWE

Water treatment equipment also manufactured by:

PHILIP HARRIS SCIENTIFIC
Scientific Supplies Co.
618 Western Avenue
Park Royal
LONDON W3 0TE
UK

GALLENKAMP EXPRESS
Belton Road West
Loughborough
Leicestershire LE11 0TR
UK

UTIL S.A
Rua 17 Febrero 156
Bonsucesso
21040 SAO PAULO SP
BRAZIL

PENDAR ENVIRONMENTAL
P.O. Box 22
Hampshire Industrial Estate
BRIDGWATER TA6 3NT
UK

WATER ENGINEERING (PVT) LTD
P.O. Box 8507
Belmont
BULAWAYO
ZIMBABWE

MERCK LTD
Broom Road
Parkstone
Poole
Dorset BH12 4NN
UK

67.0 Winnowers

Manual winnowers

HAND-OPERATED WINNOWER UNATA 5100

This is a lightweight, easy to handle winnower which can separate seed into three outlets: whole seeds, crushed seeds, and pellicles/specks of dust. Throughput: $1m^3$ per hour, seeds ready for market. Dimensions: length $99 \times$ width $64 \times$ height 121 cm.

PRICE CODE: 2

UNION FOR APPROPRIATE TECHNICAL
ASSISTANCE
G. Van den Heuvelstraat 131
3140
RAMSEL
BELGIUM

CYCLE WINNOWER

This machine can be used for separating the grain from the chaff when a favourable breeze fails. The blades have a diameter of 120cm and are fitted on a shaft running on two ballbearing races. There are two free wheels in the gear train along with a flywheel. A seat is provided for the operator and the machine is pedal-driven. The frame is all steel, and up to four people can winnow at a time. Weight: 80kg. Throughput: 1500 to 2000kg per day.

PRICE CODE: 1

COSSUL & CO. PVT LTD
123/367 Industrial Area
Fazalgunj
Kanpur
U.P.
INDIA

ALL STEEL WINNOWER & COFFEE SIEVE

This separates the dust and shells from hulled coffee. The coffee is first fed into the hopper which is fitted with a feed regulating slide and a feed roller to distribute the coffee evenly across an air current. The mixture falls vertically through the horizontal air blast and loose husks and dust are discharged. Dust-free coffee then falls onto an oscillating sieve which has two grids; upper and bottom. The upper holds back unshelled coffee, and the bottom retains shelled coffee but discharges shell and grit. The machine uses a 1.5kW (2hp) motor. Throughput: 907kg per hour.

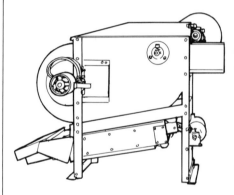

PRICE CODE: 3

JOHN GORDON & CO. (ENGINEERS) LTD
Gordon House
Bower Hill
EPPING
CM16 7AG
UK

Hand-operated winnowers currently available

Country	Manufacturer	Equipment name	Throughput (kg per hour)	Price code
Belgium	Union for Appropriate Technical Assistance	HAND-OPERATED WINNOWER UNATA 5100	$1m^3$	2
		HAND-OPERATED WINNOWER KIT 5101		2
UK	John Gordon & Co. (Engineers) Ltd	ALL STEEL WINNOWER & COFFEE SIEVE	907	3
India	Kisan Krishi Yantra Udyog	GRADER-CUM-WINNOWER	400–500	2
	Cossul & Co. Pvt Ltd	CYCLE WINNOWER	188–250	2
		HAND WINNOWER	62–100	1
Sri Lanka	Nawinne Agric. Implements Manu. Co.	HAND-OPERATED WINNOWING FAN	$0.0045m^3$	1

Powered winnowers

WINNOWER

Designed for paddy, this is a compact free-standing model, fabricated from mild steel and operated by cranking a handle, which is connected by pulley and V-belt to the fan.

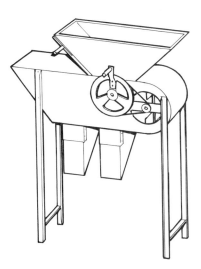

PRICE CODE: 1
COMILLA CO-OPERATIVE KARKHANA LTD
Ranir Bazar
P.O. Box 12
Comilla
BANGLADESH

CC-1 CATADOR WINNOWER FOR SAMPLES

This winnower separates the loose shells from shelled coffee quickly. Once the machine has been set into motion, material is fed into the hopper. Shelled coffee is discharged at the outlet below the hopper, whilst husks and dirt are discharged from a separate outlet. A 30W motor drives the machine. Throughput: 15–45kg per hour. Height 50cm.

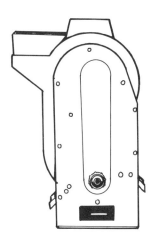

PRICE CODE: 3
JOHN GORDON & CO. (ENGINEERS) LTD
Gordon House
Bower Hill
EPPING CM16 7AG
UK

PENKHA/HUSK SEPERATOR

This machine will separate the paddy husk from dehusked rice. Both the paddy and husk are fed into a shaking sieve mounted on a husk separator. The sieve separates the broken parts and the husk powder from the rice. The machine uses an aspiration fan and is made from heavy duty steel sheets. Air control dampers, with fine adjustments, are provided for regulating the air flow. Throughput: 250–3500kg per hour. Sizes available: 61–183cm.

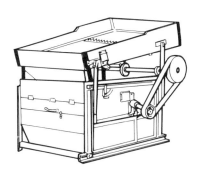

PRICE CODE: 3
DEVRAJ & CO.
Krishan Sudama Marg
Firozpur City
PUNJAB
INDIA

Powered winnowers currently available

Country	Manufacturer	Equipment name	Throughput (kg per hour)	Price code
Bangladesh	Comilla Co-operative Karkhana Ltd	WINNOWER		2
		WINNOWING MACHINE		2
India	Devraj & Company	PENKHA/HUSK SEPARATOR	250–3500	3
		DEVRAJ PADDY SEPARATOR	500–3000	1
	Dandekar Machine Works Ltd	HUSK/RICE SEPARATOR		2
Japan	Cecoco	HAND GRAIN WINNOWER	650	1
Nigeria	Department of Agricultural Engineering	MULTI-GRAIN WINNOWER	466	1
Peru	Herrandina	GRAIN WINNOWER (AVENTADORA PARA GRANOS)	200–300	1
Sri Lanka	Mahaweli Agro-Mech Industrial Complex	WINNOWING FAN		1
UK	Sathyawadi Motors & Transporters	WINNOWING FAN		1
	Blair Engineering Ltd	HAND WINNOWER AND GRAIN CLEANER	1000	1
	John Gordon & Co. (Engineers) Ltd	CC-1 CATADOR WINNOWER FOR SAMPLES	15–45	2
Zimbabwe	G North & Son (Pvt) Ltd	WINNOWER	3 bags/h	2

MULTI-GRAIN WINNOWER

This prototype model uses a centrifugal blower fan to separate the grain from the chaff. The improved model works at about 99% winnowing efficiency. An integral motor is included and the machine is free-standing. Throughput: 466kg per hour.

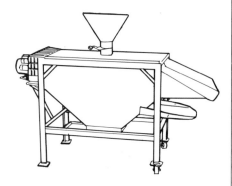

PRICE CODE: 3

DEPARTMENT OF AGRICULTURAL ENG.
Faculty of Engineering
University of Nigeria
Nsukka
NIGERIA

68.0 Miscellaneous equipment

SOYA MILK EQUIPMENT

This stainless steel equipment is designed for the production of soya milk. From one 60kg bag of soya, the machine can produce 750 litres of milk. Heating is by coiled elements in a stainless steel tank. Other equipment such as a separator, cooler and packaging machines can be connected. Throughput : 62.5 litres per hour. Dimensions: 165 × 157 × 60 cm.

PRICE CODE: 4

ORDEPAR INDUSTRIA DE MAQUINAS
Proloita Ltda
Rua Mossoro 421
86020 LONDRINA
PARANA
BRAZIL

MANUAL CAN OPENER

This has a circular blade for opening cans. The unit clamps onto work surfaces and opens cans between 2 and 40cm in height. It can be easily adjusted and the blade and the drive wheel can be easily replaced.

PRICE CODE: 2

ROSEN & ROBBERT OHHG
MASCHINENFABRIK
Hilsmannweg 23
Postfach 2230
ARNSBERG 1
D 5760
GERMANY

POPCORN MAKER

This model is designed for medium scale popcorn production. 100g of oil is heated in the machine, corn (maize) is poured in and it is heated. When 75% is roasted, the lid automatically opens slightly enabling the popped corn to fall out and the remaining corn to be popped. The temperature is controlled by thermostat. Counter and floor standing models are available. Power requirement: 220V. Capacity: 453g corn per charge.

PRICE CODE: 3

RAJAN UNIVERSAL EXPORTS (MFRS) P LTD
Raj Buildings 162
Lingi Chetty Street
P. Bag 250
Madras 600 001
INDIA

CORN POPPER, ROTARY TYPE

This machine is designed for small scale popcorn production. It has a concave bottom that keeps the unpopped corn directly over the heat. It also has a contoured propellor which sweeps the base to roll the kernels and prevent them scorching. Capacity: 3.4 lires.

PRICE CODE: 1

LEHMAN HARDWARE & APPLIANCES INC.
P.O. Box 41
4779 Kidron Road
Kidron
OHIO 44636
USA

SUGAR COTTON MACHINE

This machine produces spun sugar (candy floss). It consists of a revolving stainless steel drum within which a fine nozzle produces the strands of sugar.

PRICE CODE: 2

RAJAN UNIVERSAL EXPORTS (MFRS) P LTD
Raj Buildings 162
Linghi Chetty Street
P. Bag 250
MADRAS 600 001
INDIA

MOBILE COOLING RACK

This rack is for cooling baked foods. Dimensions: 45.7 × 35.6 × 152.4cm.

PRICE CODE: 1

MR FABRICATION AND SERVICING CORP.
Carmelo & Bauermann Compound
EDSA
Guadalupe
Makati
Metro Manila
PHILIPPINES

ICE MAKING EQUIPMENT

ICE CUBE MACHINE SD SERIES:
SD22-SD60
This machine has inner components made from plastic. The system consists of an hermetically sealed compressor with a large condensor (air- or water-cooled). The ice production cycle is fully automatic and is controlled by a timer. Capacity: 1250 or 3250 ice cubes/day. Dimensions : Weight 32 × Depth 27 × Height 56cm

PRICE CODE: 2

ICE FLAKERS

This produces dry, compact ice flakes. Throughput: 150 – 155 kg per 24 hours. Dimensions: weight 104.5 × depth 55 × height 109.5cm.

PRICE CODE: 4

PORKKA (UK) LTD
Unit 5, Olds Approach
Tolpits Lane
RICKMANSWORTH
WD1 8TD
UK

Index of manufacturers

ANTIGUA

MINISTRY OF AGRICULTURE, *136*
Dunbars
ANTIGUA

ARGENTINA

ALUMCRAFT ARGENTINA S.C.A., *83, 114*
Mercedes 568
1407 Buenos Aires
ARGENTINA

AUSTRALIA

JOHN L. GUILFOYLE (SALES) LTD, *80, 81*
P.O. Box 18
Darra
Brisbane
QLD 4076
AUSTRALIA

INLAND SALES AND SERVICE, *85*
11–13 Railway Parade
Merredin
WA 6415
AUSTRALIA

PENDER BEEKEEPING SUPPLIES PTY LTD,
80, 81, 88, 104
17 Gardiner Street
Rutherford
NSW 2320
AUSTRALIA

AUSTRIA

PUFF, STEFAN GmbH, *80, 81, 88*
Neuholdaugasse 36
8011 Graz
AUSTRIA

BANGLADESH

**COMILLA CO-OPERATIVE KARKHANA
LTD,** *147*
Ranir Bazar
P.O. Box 12
Comilla
BANGLADESH

BELGIUM

ATELIERS ALBERT & CO. S.A., *91, 115*
Rue Riverre 2
B-5750 Floreffe
BELGIUM

**UNION FOR APPROPRIATE TECHNICAL
ASSISTANCE,** *90, 91, 92, 114, 128, 133,
146*
G. Van den Heuvelstraat 131
3140 Ramsel
BELGIUM

BOLIVIA

FERROSTAL BOLIVIA LTDA, *124*
Av Mariscal Santa Cruz
Edif Hansa Piso 12 of 1 Cassila
La Paz
BOLIVIA

BOTSWANA

**RURAL INDUSTRIES INNOVATION
CENTRE,** *93, 122, 123, 145*
Private Bag 11
Kanye
BOTSWANA

BRAZIL

BRAS HOLANDA, *101, 124*
P.O. Box 1250
Curitiba
PR-80.001
BRAZIL

**BRAXINOX BRASIL EQUIPAMENTOS
INDUSTRIAIS LTDA,** *105, 109*
Estrada da Servido 155 Osasco
Caixa Postal 280
06170 Sao Paolo SP
BRAZIL

**CIMAG (COM E IND DE MAQUINAS
AGRICOLA),** *114*
Rua St Terezinha 1381
13970 Itapira SP
BRAZIL

CROYDON IND DE MAQS LTDA, *105, 108,
111, 122, 136*
Soci Gerente
R 24 de Fevereio 71 f75-77-79
Rio de Janeiro RJ 21040
BRAZIL

ECICASA INDUSTRIA E COMERCIO LTDA,
102
Rua Guaranesia 900
Sao Paulo
BRAZIL

EVANDRO FREIRE, *136*
Centro de Pesquisas do cacao
Divisao de Technologica Agricola
LX.P7
Itabuna BA
BRAZIL

INCOBI, *93*
14240 Cajuru
Sao Paulo
BRAZIL

IND DE FORNOS PERFECTA LTDA, *122*
Rua dos Ciclames 460
03164 Sao Paolo SP
BRAZIL

INDUSTRIAS MAQUINA D'ANDREA SA,
85, 107, 115
Rue General Jardim 645
Sao Paolo 01223
BRAZIL

JR ARAUJO IND E COM DE MAQS LTDA,
113, 138
Division Internacional
Rua General Jardim 645
Sao Paolo SP 01223
BRAZIL

MESSRS LAREDO, *92*
Industria e Comércio
Rua 1 de Agosto
Bauro SP 17 100
BRAZIL

MAQUINAS DE KREPS SUISSO, *108*
Rua Oscar Fredre 1784
01426 Sao Paolo SP
BRAZIL

**MENEGOTTI – METALURGICA ERWINO
MENEGOTTI LTDA,** *114*
Rua 590 Erwino Menegotti 381
89250 Jaragua do Sul
BRAZIL

**ORDEPAR INDUSTRIA DE MAQUINAS
PROLOITA LTDA,** *148*
Rua Mossoro 421
86020 Londrina
Parana
BRAZIL

SAVEIRO COMERCIO E INDUSTRIA, *105*
Rua Nicolau Ancona Lopes 153
01522 Sao Paulo SP
BRAZIL

SUPREMA EQUIP PARA IND DE PANIFICA,
89, 120, 122
Estrada Municipal SMR 340
N 532/600 Jardim Boa Vista
Sumare SP
BRAZIL

UTIL S.A., *146*
Rua 17 Febrero 156
Bonsucesso
21040 Sao Paolo SP
BRAZIL

CAMEROON

OUTILS POUR LES COMMUNAUTES, *115,
133*
BP 5946
Douala
Akwa
CAMEROON

CANADA

O'HARA MANUFACTURING LTD, *128*
65 Skagway Avenue
Toronto
M1M 3TD
CANADA

PATTERSON INDUSTRIES (CANADA) LTD,
118
250 Danforth Road
Scarborough
Ontario M1L 3X4
CANADA

Index of manufacturers

CHILE

HAHN & CO, *124*
Santo Domingo 588
Santiago
CHILE

CHINA

AFFILIATED FACTORY JIANGSU, *127*
Institute of Technology
East Suburb
Zhenjiang
CHINA

**GUANGZHOU ELECTRIC RICE COOKER
    FACTORY,** *109*
Lianhe Suburbs
North Guangzhou
CHINA

JINAN AGRICULTURAL MACHINERY, *130*
9 Tuwu Road
Shangong
CHINA

**NATONG COUNTRY PACKING MACHINE
    FACTORY,** *128*
33 Tongyang Naulu
Pingchao Town
Natong County
Jiangsu
CHINA

**RED STAR MACHINE WORKS OF
    JIANGXI,** *113*
Bong Xiang Si
Jiangxi
CHINA

TECHNOLOGY DEVELOPMENT OFFICE, *97*
S & T Comm'n of Liaoning Province
1–2 San Hao Street
He Ping District
Shenyang
CHINA

YULING MACHINE WORKS, *135*
Jianshelu
Guxian County
Henan Province
CHINA

YUNAN GENERAL MACHINE WORKS, *99*
24 Dadi Lu, Dunchang Zheng
Yunan County
Guangdong Province
CHINA

COLOMBIA

PENAGOS HERMANOS & CIA LTDA, *92,
    113, 115, 138*
Apartado Aereo 689
Bucaramanga
COLOMBIA

COSTA RICA

UNIVERSIDAD DE COSTA RICA, *145*
COSTA RICA

DENMARK

GERSTENBERG OG AGGER A/S, *99*
Frydendalsvej 19
1809 Fredersiksberg
DENMARK

MASKINFABRIKKEN SKIOLD SAEBY, *113,
    115*
Kjeldgaardsvej
P.O. Box 143
Saeby
DK 9300
DENMARK

PRESIDENT MOLLERMASKINER A/S, *113,
    115*
Springstrup
Box 20
DK-4300 Holbaek
DENMARK

PRIMODAN FOOD MACHINERY A/S, *99*
Oesterled 20–26
P.O. Box 177
DK-4300 Holbaek
DENMARK

SWIENTY, *80, 99, 104, 138*
Hortoftvej 16
Ragebol
Sonderborg
DK 6400
DENMARK

WODSCHOW & CO A/S, *88, 118*
Industrisvinget 6
Postbox 10
2605 Broendby
DENMARK

FRANCE

GERICO FRANCE, *137, 138*
25 Rue d'Artois
Paris 75008
FRANCE

**GROUPE DE RECHERCHE & D'ECHANGES
    TECHNOLOGIQUES,** *102*
213 Rue Lafayette
Paris 75010
FRANCE

ROBOT COUPE, *90, 137*
10 Rue Charles Delescuze
BP 135
Bagnolet 93170
FRANCE

GERMANY

KRAEMER UND GREBE GmbH & CO., *84,
    128*
Postfach 2149 & 2160
D-3560 Biedenkopf
Wallall
GERMANY

IBG MONFORTS & REINERS GmbH & CO.,
    88, 98, 134
Postfach 20 087 53
Monchengladbach 2
D-4050
GERMANY

PROBST & CLASS GmbH & CO. KG, *116*
Postfach 2053
Industriestrasse 28
Rastatt
7550
GERMANY

**ROSEN & ROBBERT OHHG
    MASCHINENFABRIK,** *148*
Hilsmannweg 23
Postfach 2230
Arnsberg 1
D-5760
GERMANY

SOLLICH GmbH & CO. KG, *97*
D-4902
Bad Salzuflen
GERMANY

GHANA

AGRICULTURAL ENGINEERS LTD, *91, 92,
    98, 102, 103, 106, 113, 132, 134, 136,
    137, 138, 140, 145*
Ring Road West Industrial Area
P.O. Box 12127
Accra-North
GHANA

**NATIONAL COUNCIL ON WOMEN AND
    DEVELOPMENT,** *140*
P.O. Box M53
Accra
GHANA

TECHNOLOGY CONSULTANCY CENTRE,
    132, 133
University of Science and Technology
Kumasi
GHANA

HUNGARY

**SALINA GEPGYARTO ES
    TOMITESTECHNIKA,** *132*
Vallat
H-2490 Poszfaszblocs
Iskola V.1
PF10
HUNGARY

INDIA

ANJALI KITCHEN COMPONENTS, *98*
Mahavir Brothers
202 Trivani House
Kelivadi Back Road
Bombay 400 004
INDIA

ASHOK TRADING CO., *139*
18 R N Mukherjee Road
Calcutta 700 001
INDIA

BEHERE'S & UNION INDUSTRIAL WORKS,
    86, 104, 133
Jeevan Prakash
Masoli
Dahanu Road 401 602
Dist Thane
INDIA

BHARAT INDUSTRIAL CORPORATION,
    102
Petit Compound, Nana Chowk
Grant Road
Bombay 400 007
INDIA

BIJOY ENGINEERS, *89, 103, 120, 122*
Mini Industrial Estate
P.O. Arimpur, Trichur District
Kerala 680 620
INDIA

B.M. ENGINEERING WORKS, *137, 138*
PB No. 12
Naidu Street
Chikmagalur 577 101
INDIA

Index of manufacturers

SCREENIVASA ENGINEERING WORKS, *116, 125*
2-1-460/1
Mallakunta
Hyderabad 500 044
INDIA

SHACO ENTERPRISES, *125*
161/163 R R Mohan Roy Road
Pharthana Samaj
Bombay 400 004
INDIA

SHREE RAJALAKSHMI INDUSTRIES, *139*
No. 61 13th Cross
J.P. Nagar 3rd Phase
Bangalore 560 002
INDIA

SQUARE TEC, *102*
35 Industrial Estate
Lonavla 410 401
INDIA

STADLER CORPORATION, *101, 117, 123*
P.O. Box 7177
Bombay 400 070
INDIA

STANDARD AGRICULTURAL ENG. CO., *145*
824/5 Industrial Area B
Ludhiana
Punjab 141 003
INDIA

THAKAR EQUIPMENT COMPANY, *90, 105, 107, 108, 109, 123, 128, 131*
66 Okhla Industrial Estate
New Delhi 110 020
INDIA

TINYTECH PLANTS PRIVATE LTD, *98*
Near Mallavia Wadi
Gandal Road
Rajkot - 360 002
INDIA

WATRION WATER & FILTER ENGINEERING P LTD, *146*
34A Rameshwar
4 Bunalow Road, Andhari (W)
Bombay 400 058
INDIA

WONDERPACK INDUSTRIES PVT LTD, *127*
72, 1st Floor, Shivial Mansion
Lamington Road
Bombay 400 008
INDIA

INDONESIA

CV KARYA HIDUP SENTOSA, *86, 97, 114*
JL Magelang 144
Yogyakarta 55241
INDONESIA

KEMAJUAN, *137*
Irian Jaya 3
Malang 63118
East Java
INDONESIA

PABRIK MESIN, *98*
P.T. Mata Hari SS
JL Cempala 18
INDONESIA

PT RUTAN MACHINERY TRADING CO., *97, 145*
P.O. Box 319
Surabaya 60271
INDONESIA

ITALY

LA PARMIGIANA, *129, 130*
GB Della Chiesa 10
Fidenza (Pr) 43036
ITALY

JAPAN

CECOCO, *91, 92, 93, 97, 98, 112, 114, 145, 147*
P.O. Box 8
Ibaraki City
Osaka 567
JAPAN

KENYA

DATINI MERCANTILE LTD, *141*
Enterprise Road
Box 45483
Nairobi
KENYA

ILO – SDSR, *86, 95, 106, 141*
P.O. Box 60598
Nairobi
KENYA

KOREA

KOREA FARM MACHINE & TOOL INDUSTRIAL CO-OP, *86*
11–11 Dongja-Dong
Youngsan-Gu
Seoul
KOREA

LIBERIA

AGRO MACHINERY LTD, *91, 112, 114, 115, 134, 137, 145*
P.O. Box 3281
Bush Rod Island
Monrovia
LIBERIA

MALAWI

MINISTRY OF AGRICULTURE, *91*
MALAWI

MEXICO

DISENSOS Y MARQUINARIA JER S.A., *78, 135*
Emiliano Zapata 51
Col Buenavista 54710
Cuautitlan Itzcalli
MEXICO

EQUIPOS HERGO INTEGRALES, *118*
AP Division del Norte 2984
C.P. 04370
MEXICO DF

INDUSTRIAL DE PARTES S.A. DE C.V., *79*
Latoneros Num. 99
Col Trabajadores del Hierro
MEXICO DF 02650

MAPISA INTERNATIONAL S.A. DE C.V., *84–5*
Eje 5 Oriente Rojo Gomez 424
Col Agricola Oriental
MEXICO DF 08500

MAQ INDUSTRIAL PARA LA PULVERIZACI, *114*
Ruben M Campos 2762
Col Villa de Cortes
03530 MEXICO DF

MAQUINARIA OVERENA S.A. DE C.V., *88, 103*
Av. Division del Norte 2894
Col Parque San Andres
MEXICO DF

PRACTIMIX S.A., *118*
Oriente 83 No 4311
Col Malinche
MEXICO DF C P 07880

NETHERLANDS

GEBR. RADEMAKER, *77, 80, 81, 82, 84, 86, 87, 112, 120, 141, 144*
P.O. Box 81
3640 AB Mijdecht
NETHERLANDS

ROYAL TROPICAL INSTITUTE, *133*
Mauritskade
1092 AD Amsterdam
NETHERLANDS

SPANGENBERG BV, *88, 89, 90, 117, 118, 137*
Grootkeuken Professionals
De Limiet 26
4124 PG Vianen
NETHERLANDS

VOTEX TROPICAL, *145*
Vogelenzang Andelst BV
Wageningsestraat 30
Andelst
NL-6673DD
NETHERLANDS

NEW ZEALAND

MACKIES PTY LTD, *122*
Unit A, 5 Lorien Place
P.O. Box 82034 Highland Park
Auckland
NEW ZEALAND

STUART ECROYD BEE SUPPLIES, *88*
P.O. Box 5058
Papanui
Christchurch 5
NEW ZEALAND

NIGERIA

AGRICULTURE ENGINEERING DEPARTMENT, *103, 116*
University of Ife
NIGERIA

DEPARTMENT OF AGRICULTURAL ENGINEERING, *96, 105, 115, 134, 147, 148*
Faculty of Engineering
University of Nigeria
Nsukka
NIGERIA

INSTITUTE FOR AGRICULTURAL RESEARCH, *91, 92, 96, 112, 145*
Samaru, Ahmadu Bello Uni
PMB 1044
Zaria
NIGERIA

M/S DALTRADE (NIG) LTD, *96*
Plot 45 Chalawa Indust Est
P.O. Box 377
Kano
NIGERIA

Index of manufacturers

Index of manufacturers

CS BELL CO., *112, 114*
170 W Davis Street
Box 291
Tiffin OH 44883
USA

DADANT & SONS INC., *80, 85*
Hamilton
Il 62341
USA

FMC CORPORATION, *76, 79, 89, 99, 100,*
112, 126, 129, 131, 136
Machinery International Division
P.O. Box 1178
San Jose
California
95108
USA

FOOD PRESERVATION SYSTEMS, *137*
1604 Old New Windsor Road
New Windsor
Maryland
21776
USA

GARDEN WAY SQUEEZO, *137*
1 Mill Street
Burlington
VT 05401
USA

LEHMAN HARDWARE & APPLIANCES
INC, *78, 80, 81, 82, 83, 84, 86, 87, 89, 91,*
95, 97, 99, 105, 107, 108, 110, 111, 112,
113, 120, 121, 127, 130, 131, 132, 135,
136, 142, 144, 148
P.O. Box 41
4779 Kidron Road
Kidron
Ohio 44636
USA

RR MILL INC., *79, 83, 91, 112, 113, 114,*
115, 130, 135, 136, 137
45 West First North
Smithfield
Utah 84335
USA

VITANTONIO, *108, 120, 121, 130*
Dept C
34355 Vokes Dr
Eastlake
Ohio 44094
USA

URUGUAY

FRAGOL LTDA, *101, 124*
Av Libertador Brig Gral
Lavelleja 1641-Esc 203
Montevideo
URUGUAY

YUGOSLAVIA

RUDARSKO METALURSKI KOMBINAT,
114, 132
Ulica 29
Novembra br. 15
Kostajnica
79224 Bos
YUGOSLAVIA

ZAMBIA

CENTRAL FISHERIES RESEARCH
INSTITUTE, *140*
ZAMBIA

ZIMBABWE

AVERY ZIMBABWE PVT LTD, *143*
P.O. Box 392
Harare
ZIMBABWE

HC BELL & SON (PVT) LTD, *115, 116*
Mutare
ZIMBABWE

ENDA ZIMBABWE (PVT) LTD, *92*
14 Belvedere Road
P.O. Box 3492
Harare
ZIMBABWE

GDI (PVT) LTD, *106*
P.O. Box 2202
Harare
ZIMBABWE

GRUENTHAL & BEKKER (PVT) LTD, *109*
P.O. Box 2550
Harare
ZIMBABWE

MITCHELL & JOHNSON PVT LTD, *106,*
109, 117, 123, 128, 131, 139
P.O. Box 96
Bulawayo
ZIMBABWE

MULTI SPRAY SYSTEMS, *91*
P.O. Box HG 570
Highlands
Harare
ZIMBABWE

G. NORTH & SON (PVT) LTD, *91, 112, 147*
P.O. Box BPX ST 111
Southerton
Harare
ZIMBABWE

PLAZA ORIENTAL (PVT) LTD, *106*
72 Manica Road
Harare
ZIMBABWE

QUICKFREEZE (PVT) LTD, *86*
P.O. Box 368
Southerton
Harare
ZIMBABWE

STAINLESS STEEL INDUSTRIES PVT LTD,
106, 129
Hermes Road
P.O. Box ST234
Southerton
Harare
ZIMBABWE

T.A. INDUSTRIAL DIVISION, *86*
P.O. Box 3910
Harare
ZIMBABWE

TONY ELECTRICAL MANUFACTURING, *114*
688 Cooleen Road
New Ardbennie
Harare
ZIMBABWE

WATER ASSOCIATES & EQUIPMENT, *146*
P.O. Box HG 423
Highlands
Harare
ZIMBABWE

WATER ENGINEERING (PVT) LTD, *146*
P.O. Box 8507
Belmont
Bulawayo
ZIMBABWE

Index of equipment

Index of equipment

Questionnaire

Please help us to improve future editions of *Small-Scale Food Processing* by completing this questionnaire and sending to: Small-Scale Food Processing, ITDG, Myson House, Railway Terrace, Rugby CV21 3HT, UK.

Please tick box(es) where appropriate (tick more than one box if applicable).

Name ....

Position/profession ....

Address ....

....

....

....

User/organization:

Commercial company ☐		Academic institution ☐
Individual ☐		Government organization ☐
International non-govt organization ☐		National non-govt organization ☐
Other (please specify) ☐		....

How did you find out about this book?

	Word of mouth ☐	Advertisement ☐
Saw it in a bookshop/library ☐	*Books by Post* ☐	IT Publications trade list ☐
Referred to in another book ☐	Other (please specify) ☐	....

For what purpose did you buy this book?

To find manufacturers'/suppliers' address ☐

To discover the range of equipment available ☐   To find information on small-scale food processing ☐

Other (please specify) ☐   ....

Which chapter(s) are of most interest to you? ....

Which chapter(s) are of least interest to you? ....

If you have any other comments please write them on the back of this questionnaire.